UNEARTHING CULTURALLY RESPONSIVE MATHEMATICS TEACHING

The Legacy of Gloria Jean Merriex

Emily P. Bonner

Hamilton Books
A member of
The Rowman & Littlefield Publishing Group
Lanham · Boulder · New York · Toronto · Plymouth, UK

Hamilton Books
4501 Forbes Boulevard
Suite 200
Lanham, Maryland 20706
Hamilton Books Acquisitions Department (301) 459-3366

Estover Road
Plymouth PL6 7PY
United Kingdom

Printed in the United States of America
British Library Cataloging in Publication Information Available

Library of Congress Control Number: 2010936056
ISBN: 978-0-7618-5399-2 (paperback : alk. paper)
eISBN: 978-0-7618-5400-5

∞™ The paper used in this publication meets the minimum requirements of American National Standard for Information Sciences—Permanence of Paper for Printed Library Materials, ANSI Z39.48-1992

Dedicated to Gloria Jean Merriex

Contents

Figures

Foreword

Dr. Thomasenia Lott Adams
Mathematics Educator, University of Florida

As I read this book, the word "transformation" kept coming to mind. From the book's conception to the last sentence, it exudes transformation. Transformation is most readily evident from observing Emily while she was in the process of writing her book. I was a witness to her most triumphant moments as well as many challenges she faced (some expected, some unexpected, some practical, some emotional) as she immersed herself in the project that yielded her book.

Spending three years in the school that served as the research setting positioned Emily to not only grow personally and professionally, but to also come face to face with her misunderstandings and misconceptions about teaching and learning among African American children. These initial misunderstandings and misconceptions stemmed primarily from a lack of exposure to and awareness of African American culture. As such, Emily began this work with questions such as, "How will I relate to these children?"; "How will I relate to this teacher?"; "What am I supposed to see happening?" and "What kinds of questions can I ask, should I ask, do I have the courage to ask?" However, in the concluding days of Emily's engagement in the school, she engaged in pondering such questions as "How can we capture the power of this teacher's craft for sharing with others?"; "What motivates these children to want success when others say that they do not want success nor can have success?"; and "What keeps the teacher-student interaction bonded in the classroom?" Emily transformed from an outsider with insecurities to a participant in the learning community with scholarly curiosities that were satisfied by classroom observations, interviews with students and interviews with the teacher. Having worked closely with Emily, I am certain that she has concluded that, because of her experience in the particular classroom, she is a more informed and more enlightened person.

Transformation happened among Emily's colleagues as well. Fellow doctoral students joined Emily in classroom observations and were astounded by her total immersion in the context and the ways in which Emily not only adapted to the unfamiliar environment but also, in essence, clothed herself in it with a sense of personal and professional fulfillment.

This book also caused personal transformation for me. One of the primary elements of my transformation came as I worked alongside Emily as I considered what it means for a context to be culturally relevant. Throughout this process, I often wondered what culturally responsive teaching meant for individuals and learning communities in terms of definition and impact. I believe that in the midst of reading various iterations of Emily's manuscript, I came to understand her perspective on this topic and realized that it was reflective of my own—that in the context of education, valuing what is meaningful to others, acknowledging this value in things we do and say, and demonstrating this understanding in the ways we act and respond in the teaching and learning moment are of utmost importance.

Finally, this book is transformative for mathematics educators. What is culturally response mathematics teaching? This is the question Emily set out to answer. It is a question that I believe she began to answer with details not previously presented in the literature but by showcasing a teacher, Gloria Merriex, that I will describe (in the words of Gloria Ladson-Billings) as a "dreamkeeper". Gloria Merriex was remarkable in how she exercised her mathematics teaching craft to engage African American children who before reaching her class were targeted as low achieving students, students of failure. Emily eloquently tells the story of this educator through her own voice and the voices of Gloria's students, colleagues, and family members. As a result, what started out as a simple doctoral student research project has turned into a message that provides clarity for those who wish to make the teaching of mathematics more involving and inviting for diverse populations of children. Emily has contributed a book about teaching and learning and learning mathematics in a diverse setting that is surely informative and definitely transformative.

Preface

"They say I was made for teaching"

-Gloria Jean Merriex

Phenomenon (Phe-nom-e-non)[1]:

1. an object or aspect known through the senses rather than by thought or intuition
2. a rare or significant fact or event

If you look inside an anatomy textbook, you will find illustrations meant to show you the depth and complexity of particular parts of the human body. For example, on one page, you will find a large picture of a hand. The next few pages are transparent overlays for the initial picture, with each showing a different level of the anatomy. The first overlay will show you the bones in the hand, with the nerves being added with the next overlay. Next comes the tissue, followed by veins and skin layers. Once all of the overlays are in place, the full complexity of the hand and all of its parts are shown. While at times it may be helpful to have the layers separated, it is when the picture is complete that the entire representation makes sense. My work in writing this monograph has been similar, with layers upon layers of a story emerging to reveal something whole and remarkable.

The foundation, or initial outline, for this work came in 2005 when I was a first year doctoral student. As part of my research assistantship, I was assigned to observe and take notes on the teaching practices of an elementary school mathematics teacher by the name of Gloria Jean Merriex. I will never forget the first time I visited her classroom—I was stunned at her innovation and deeply inspired by the stories of the children she instructed. Her mathematics teaching and its impact on her students' academic achievement piqued my initial interest, but as I came to know Gloria over the next three years, I realized that her work went much deeper and much wider than one could ever imagine. This was evidenced when Gloria was nominated by members of her community (who did not know that I even knew Gloria) to be a part of my dissertation study about highly successful mathematics teachers of African American children.

Over time, what began as a simple project meant to describe some of her mathematics teaching methods became a layered picture that culminated in significant ways. Gloria was a woman who was able to change a student's perception of himself as a learner, as an African American, and as a member of society. Starting in the classroom, the whole of her work impacted a school and a community. When she suddenly passed away in May of 2008, these transformations became particularly evident. This event pushed me to expand my work and share the practices of Gloria Merriex in an effort to pay tribute to her legacy. In speaking with her family, colleagues, administrators, students, and parents, and telling their story on the following pages, a whole, layered picture of this phenomenal woman has begun to materialize.

Though many people have a story about Gloria or her work, her personal life was somewhat private, adding to the complexity of her character. Individuals may have seen Gloria present at a conference, perform with students, or may have just heard stories about "that amazing teacher in Gainesville, Florida who used raps to teach mathematics and had remarkable standardized test scores even though she taught an underserved population of students". Gloria was talented in so many ways, but was best known for her innovative techniques in teaching mathematics and unwavering belief in children's ability to excel. Using chants, choral responses, dances, and rhythms, Gloria was able to communicate mathematical ideas to children in a way that empowered them as learners and individuals. When she was teaching, you could hear the students' voices down the hall, responding in unison.

When in her classroom, I was inspired—I almost couldn't help it. I, and others, truly did feel her teaching with your senses and not only with your mind. In fact, one of my favorite parts of working with Gloria was watching the looks on people's faces the first time they watched her engage her students. Whether an observer was a parent, professor, administrator, new student, or teacher, you could see the surprise and delight in their faces within the first five minutes. After watching her, observers would have a million questions: *How is this possible? Did they rehearse? Did she always have this many students in her classroom? Can I see more?* These reactions came about for many reasons such as the high level of engagement among students and the sheer energy that Gloria constantly exhibited. Moreover, she worked in a largely African American school in a low-income area with a population that has traditionally struggled in education.

Given her work and influence, Gloria's sudden (albeit peaceful) death is a palpable loss to her family and the surrounding community, the school, the state, and the field of education as a whole. This was painfully evident at her funeral where the mix of people in attendance was staggering. The individuals in attendance represented a range of backgrounds and had varied interests in her work and life. To some, her life, demeanor, ideas, and general techniques were of interest. To others, the intricacies of her work, its impact on mathematics education practice, and implications for teaching mathematics in poverty schools was exciting. To some, she was just Jean.

There are many reasons that I feel compelled to share the life, teaching practices, and social impacts of Gloria Merriex. Certainly, she was a master teacher who continuously went above and beyond for her students, and was passionate about their success. There was something bigger, however, on which her work made us focus. Despite the academic struggles faced by African American students, she was determined to prove that economic status and race were not central to academic excellence. She believed that learning was a matter of access, not ability, and it was up to teachers to make subjects such as mathematics, that have a reputation of being difficult, accessible to students who may have been told (verbally or otherwise) that they could not learn. At the same time, she would push these learners, who may have negative beliefs about their academic potential, to excellence, no matter who they were. She particularly focused on students who had been labeled as "learning disabled" or the like, often pushing them to realize their true ability. Often, she would guide these students from the special education classes that they had been taking into honors classes in which they excelled.

Gloria's success in the classroom reached far beyond her school. Her determination and involvement in the community resonated with the entire neighborhood, transforming and empowering those with whom she came in contact. She was able to effectively interweave her artistic talents such as singing, dancing, sewing, and cooking, into her mathematics curricula, school events, and student performances. She believed that no child (or family) should be without resources, and she worked tirelessly to provide for students who might need food, clothing, or even money. She was in touch with the parents of Duval, and the community at large. Over 30 years, she taught generations of students, and effectively helped to transform a school and a community. This work, which is far reaching and historically significant, needs to be shared.

As a teacher of mathematics, Gloria provided another level of success that existed within this larger framework. As someone who shared a childhood similar to that of many of her students, she was able to connect with learners at a deep level. Gloria used the culture of the community, which she lived in and learned about from her students, to guide her mathematics lessons and teaching techniques. She was constantly revisiting her lessons and making improvements based on student feedback. She used culturally relevant language and disciplinary tactics that helped her students to excel.

In the field of mathematics education, there are very few specific examples of teachers who teach mathematics from a culturally responsive framework. Gloria provides a subject-specific model of this construct. As such, as a researcher, there is a need to break down her specific pedagogical tools, using these ideas to conceptualize what culturally responsive mathematics teaching can look like in action. Of course, there is not (and there should not be) a formula for this type of instruction, but the field will benefit by beginning to delve into specific teaching methods that might help to address the achievement gaps that exist in mathematics while including the voices of African American teachers and community members.

In an effort to tell her story from these various points of interest, the main goal of this monograph is to deconstruct and generalize Gloria Merriex's teaching methods from a research standpoint to identify key components of her work in an effort to help the field of mathematics education to gain a more solid understanding of culturally responsive mathematics teaching. Her story is told here through her own voice and through the voices of those who knew her best. This monograph allows the reader to explore Gloria Merriex's work holistically, couched in the current state of affairs in mathematics education.

The organization of the monograph reflects this goal. Part I explores the past, focusing on the personal side of Gloria's life from childhood with an emphasis on developing several personality traits that were central to her character.

Part II zooms in more specifically on Gloria's recent life, teaching practices and pedagogical methods. An initial theory is presented that is meant to organize the complexities of her classroom and begin to generalize the key aspects of her mathematics instruction. In Part III, I look to the future, discussing outcomes and implications of Gloria's work. Student voices are included in this section as well, as they are the true testament to Gloria's methods. This section also aims to take a broader view of these ideas, tying the pieces of the theory together in ways that will help rethink key aspects of mathematics teacher preparation and mathematics instruction in general.

What you will not find in this monograph are specific instructions that dictate how to be an effective mathematics teacher in a largely African American school. Such laundry lists cannot possibly be helpful as each classroom and each teacher are different. Further, Gloria verbalized her distaste for scripted curricula on more than one occasion; it would be a dishonor to her if I then turned her methods into such a product. Rather, the theory presented is meant to give an idea of where teachers may want to focus their efforts in becoming culturally responsive educators. Particular techniques may be useful to some, while other ideas will inspire innovation and a re-thinking of the educational system.

Main Characters

Ms. Gloria Jean Merriex—"Gloria", also referred to as "Jean" by those closest to her, is the central figure of this story. She taught mathematics and reading at Duval Elementary school for 30 years, inspiring students, parents, and a community with her innovative teaching practice.

Mrs. Cenia Merriex—"Cenia" is the mother of Gloria. She and Gloria were very close, living in the same house for nearly 60 years.

Ms. Jan Merriex—"Jan" is Gloria's youngest sister. Gloria was Jan's protector when the two were growing up. Jan works with special needs children.

Dr. Leanetta McNealy—"Lee" is the principal at Duval Elementary School. Lee worked closely with Gloria, and the two traveled together to conferences, workshops, and performances all over the United States.

Dr. Buffy Bondy—"Buffy" is the current director of the School of Teaching and Learning at the University of Florida. For 9 years, Dr. Bondy was a Professor in Residence at Duval, and worked closely with the faculty, staff, and administration. She was a central figure in the school's reform efforts over the past years.

Ms. Lilliemarie Harvey—"Lilliemarie" is a 4th year teacher at Duval Elementary School. She first came to Duval in 2005 and immediately bonded with Gloria, her "Godmom." When Lilliemarie was a new teacher, Gloria served as her mentor, and the two were close at work and in their personal lives ever since.

Mr. Leonard Marshall—"Leon" is a teacher at Duval Elementary School. Currently a member of the 3rd grade team, Leon grew close with Gloria when he co-taught with her during his first year at the school. Mr. Marshall continues to teach using many of the techniques that he learned from Gloria. The two loved to sing together.

Mrs. Angela Terrell—"Angie" is the fine arts coordinator at Duval. Angie came out of retirement in 1999 to help the school infuse the arts throughout the curriculum, and has had much success in this role. Angie worked very closely with Gloria, helping her to plan and coordinate mathematics team performances and conference presentations.

Dr. Emily P. Bonner—I am the author of this work. I began observing in Gloria's classroom in 2005 and continued my work with her until her passing in 2008. I had the pleasure of interviewing those whose words drive this story.

Acknowledgements

First and foremost, I would like to acknowledge the Merriex family who showed great strength in a time of great grief. Without your candidness and honesty, this work would not have come to fruition. Gloria's legacy lives on through your words. Thank you also to the members of the Duval community who graciously offered their time, knowledge, and care during the preparation of this work, especially Dr. Lee McNealy, Dr. Buffy Bondy, Angie Terrell, Lilliemarie Harvey, and Leon Marshall. I hope I have represented your words well. I would also like to thank Dr. Thomasenia Lott Adams, my guide and mentor (and editor!) through the process of completing this manuscript, and Dr. Don Pemberton and the U.F. Lastinger Center for your unwavering support. Lastly, I would like to acknowledge Gloria Jean Merriex. The impact of her instruction lives on in the success of her children and students.

Part I: The Life of a Master Teacher

Immediately after emancipation, Black educators assumed the unique task of enhancing opportunities for newly freed slaves. These racial uplift teachers, mostly women, taught in segregated schools to prepare Black children for freedom, respectability, independence, and self-reliance. This same tradition of Black teachers as racial uplift professionals continued and thrived in segregated schools, particularly in the South. A key lesson to be derived from this research is how the oppressive circumstance of segregation resulted in a functional, semiautonomous Black community with its own peculiar set of rules, norms, sanctions, and rewards.

-Jacqueline Jordan Irvine, *Educating Teachers for Diversity: Seeing with a Cultural Eye, 2003, p.56*

Chapter 1

Society might say that you can't learn because of your social status. Low socioeconomic, free lunch, east side, so they think you can't learn, but that's not true, and it's up to you to show that you can learn. I can understand it because of where I come from. I tell them about my house where I used to live, and I say, look I'm in front of you now, so ***it can be done****...One part of town does not want to even come over here with you because they figured that you couldn't learn. But that's not true, it's a given fact that you can learn, but show them that you can.*

-Gloria Jean Merriex

The city of Gainesville, Florida is located in north central Florida and is the largest city in the county of Alachua. Placed between the panhandle and the tail of Florida, the city is roughly 80 miles southwest of Jacksonville and about 140 miles northeast of Tampa. More rural than urban, the population currently hovers around 114,000, with close to 17 percent of families and 34 percent of individuals living below the poverty line[1].

Though the city is relatively small, about 49 square miles in area, there are distinct cultural influences in various parts of the city. Demographically, the city is about 67 percent white, 24 percent African American, and 8 percent Hispanic; these populations are somewhat segregated, largely residing in different parts of the city. The west side of Gainesville houses the University of Florida and most of the white residential population with a modest amount of diversity, while the east side of the city is home to several mostly African American communities. This distinction is evident not only in census numbers, but also in school populations, church congregations, and neighborhood families. In fact, it was not until several years after 1954's famous *Brown vs. Board of Education* decision that Gainesville public schools were desegregated. Despite years of integration, east side schools are still predominately populated by African American students.

In the years of official segregation, Lincoln High School (which actually housed seventh through twelfth grade students) served the east side of town's African American middle and high school population, while Gainesville High School, which is located on the west side of town, served white students. In 1956, Lincoln High School was moved to a new building in southeast Gainesville, and the old campus came to house A. Quinn Jones Elementary School, which was also populated by young, black students.

After some restructuring over the years and the introduction of middle schools, Lincoln now exclusively serves students in grades 6-8, with Eastside High School serving high school students in the area. There are several other elementary schools and one other middle school in east Gainesville, and A. Quinn Jones is now a Center for Exceptional Students which serves severely emotionally disturbed students. Though schools on the east side of town have some ethnic diversity, many still have populations that are overwhelmingly African American[2].

Duval Elementary School, located in the northeast portion of Gainesville in a community known as Duval Heights, is just a few miles from the current Lincoln Middle School campus. Duval Heights is almost exclusively African American. There is a sense of community in the area that is evident as you drive down the main road. The many times I drove through the neighborhood to the school, I often saw people riding their bikes to and from the convenience store, friends conversing outside of one of the many churches, children walking home from school together, and friends sitting on their porch chatting; there were always at people outside. The homes are quaint and older with colorful character and unique qualities, and many are shaded by the large trees that line the streets.

About a minute off of the main turn into Duval Heights, after having passed numerous churches sitting on the avenue, is Duval Elementary School. The campus is modest and older, but not unkempt. This creates a sense of community that is evident the moment you walk through the doors. The front desk staff is always friendly, often chatting with parents or teachers while helping visitors find their way. This same kind demeanor is consistent throughout the school's administration, faculty, and staff. The walls of the school are decorated with student work and newspaper clippings that document the accolades that the school has received. Student work consistently lines the hallways outside of the main classroom corridor.

The school draws most of its students from this neighborhood despite a district-wide fine arts magnet program and recent success on state-mandated tests such as the Florida Comprehensive Assessment Test (FCAT). Like Lincoln, the school's population is almost exclusively (about 98 percent) African American, with roughly 85 percent of students qualifying for free or reduced-lunch programs in 2008[3] making it a Title I school. Despite past struggles, over each of the past 6 years, the school has earned 4 A's (in 2003, 2004, 2006, and 2008) and 2 B's (in 2005 and 2007)[4] as measured by the state of Florida accountability system, continually showing adequate yearly[5] progress in most areas.

Duval's recent successes in 5th grade mathematics scores and gains was due in large part to Gloria Jean Merriex, a mathematics teacher who spent thirty years as a faculty member at the school. A native of Gainesville, Gloria was a product of the city's ongoing race struggles that were reflected in the continued segregation of schools. Resulting experiences and others drove her belief in children, and historical cultural influences drove her teaching methods. What was exceptional about Gloria, however, was much more than her teaching. As this story will show, Gloria's work contributed to the transformation of a neighborhood, the unparalleled success of a school, and the empowerment of generations of learners.

The bustling, pink home of the Merriex family sits in the east part of town less than one mile from Duval Elementary and less than 3 miles from the University of Florida. Here, Cenia and Donal Merriex Sr. raised their five children. "We didn't have much," Jan, their youngest daughter, recalls, "but everything we did was family . . . my daddy instilled that in us." Jan and her siblings, Donal Jr., Gloria, Mary, and Charles, have carried this value throughout their lives as became obvious when I began speaking with Jan, Charles, and Cenia.

Gloria, the second child, entered the world on November 16, 1949—one day after her father's birthday, and one day before her mother's—at 7 pounds, 2 ounces. Gloria was particularly close to her mother, Cenia, especially later in life. "They were like two peas in a pod," Jan says. "They had that same forceful personality." Gloria and Cenia lived together in the Merriex home for 55 of the last 58 years. "Ever since [Gloria has] been in the world she's been here...she was right here until the time she went off to school. When she got through with school she was right back here in the house," Cenia says, "until the day she died." Gloria was also very close to her father until he passed. "She was like a daddy's child," Cenia says, "but don't nobody miss her like her mama."

As a child, Gloria developed the strong personality that would make her so successful later in life. "She would protect me at all costs because I'm the youngest," Jan recalls. "When we were growing up the older ones would have to take care of the younger ones, so no matter what was going on they protected us. She would still do that even now that I am older!"

Gloria would also shield Jan from the realities of the day. When Jan was a little girl, Gloria worked at Videll's Drugstore. Early on Saturday mornings, Gloria would wake Jan and together they would walk to Videll's to earn a bit of money. When they arrived at the downtown shop, Gloria would always direct Jan to the back door, which was often locked. This confused Jan, as the front door was always open. "Why do we have to go this way? This door is locked so why don't we just go around front?" she would ask Gloria. "I didn't recognize because I was still young that...back in those days...blacks go to the back," but Gloria would never directly say that to young Jan; rather, she would just wait for the door to be opened. She would never tell me "oh, you can't do that because . . . blacks can't come around on the front," Jan recalls. Gloria was well aware of such social rules, having attended A. Quinn Jones Elementary and Lincoln High School and experiencing segregation firsthand.

The experiences that Gloria had while segregation was still officially intact, and the attitudes she developed because of those experiences would later be one key to her success in the classroom. This type of experience strongly informing teacher practice is not uncommon, though it presents several interesting moral contradictions. African American educators who have experienced the realities of formal segregation are somewhat rare because of age, and are often underrepresented in terms of voice and influence. Given their life history, however, these teachers are in unique positions to connect with African American children who might be affected by similar cultural struggles today which are brought on by overt and systemic prejudice and racism. These teachers are often intricately involved with students not only in this context of academics, but also in guiding learning about life and the world, and the dangers that faced them outside of school. This is particularly important because of these teachers' own involvement in segregation, unique perception of racism, and experiences within an oppressive societal and educational system.

These educators "bring diverse family histories, value orientations, and experiences to students in the classroom, attributes often not found in textbooks" (Pang & Gibson 2001, 260). As such, these educators were in a uniquely powerful position to influence student beliefs about self, racial identity development (Tatum 1997), and to teach authentic content through which students were able to better understand the political and social situation of the country (Freire 1998). Interestingly, while it is often recognized that separation by race creates an inherently unequal arrangement (due to the magnitude of other factors in the educational system), much of this work implies that before integration, some African American students were in academic contexts that provided more access to academic and cultural knowledge (Ladson-Billings 2004).

In separate settings where students could learn from the struggles of African American teachers and familiar role models in edifying ways, learners were better understood in terms of cultural behavior, were not forced to conform to Eurocentric values, and were provided culturally relevant instruction. This is not meant to imply that desegregation was a bad idea, but is meant to highlight the fact that it was not the schools or teachers in black schools that caused an inequitable situation. Black students may have actually received a better education in separate schools where there were teachers of their race (Siddle-Walker 1996). It was the "larger system that defined [students] and their schools as inferior" (Ladson-Billings 2004, 5) and the perceptions created by that system that made separate schools inherently unequal.

Many African American teachers lost their jobs when integration occurred, and the decline of teachers of color has been consistent ever since (Hudson & Holmes 1994). Today this trend continues; there is a severe shortage of teachers of color in the teaching force (Howard 2006) and in teacher preparation programs. In 2000, of 3 million public school teachers, roughly 9 percent of those teachers were African American (Jorgensen 2000; Snyder 1999), and 87 percent were white while students of color constitute about 40 percent of the student population (The American Association of Colleges for Teacher Education [AACTE] 1999). Many schools (over 40 percent) do not employ any teachers of color, and "many students will complete their schooling without being taught by a single teacher of color" (Irvine 2003, 52). As a result, many students do not have the benefit of these types of role models in school. Teacher preparation programs show little hope for curbing this trend, as white females constituted over 80 percent of the population in colleges of education in 1999 while African Americans represented only 9 percent of prospective teachers in such programs (Yasin 1999).

As a result of the white dominance in the field, much of the research and literature in education, and particularly in mathematics education, largely leave out the voices of African American teachers (Agee 2004; Hooks 1994) despite their distinctive and culturally based pedagogies and management strategies. As such, educators (and ultimately students) are missing perspectives and worldviews of these women, and curricula, dominant pedagogies, and teacher preparation programs (Gay 2000) reflect this void; many teachers are compelled to teach specific material in particular, non-cultural ways.

Accordingly, African American students often do not believe that teachers understand their sociopolitical struggles. As Irvine (2003) urges, "we should publicize success stories of teachers of color widely" (60). Gloria is certainly one of these success stories who bridged cultural gaps in mathematics by using culturally based teaching methods and familiar communication patterns. Moreover, having lived as an African American through decades of blatant and invisible inequities in all parts of life, Gloria brought a realistic and understanding frame of reference to her students. This historical context allowed her to educate students not only in mathematics, but also in strength of character and beliefs about self.

Gloria achieved greatness through her own path of self-discovery. She continually challenged her place in society and worked to create opportunities for herself. After graduating from Lincoln High in 1967, she worked in the cafeteria of a local school as a cook. Her brother Charles remembers that "[Gloria] would say, 'I've got to do something better than this.' She didn't want to stay [cooking at the school]. She knew it wasn't for her." Despite financial barriers, Gloria found a way to change her life by going to college. Cenia recalls, "We did not have money to send our children to school, but [Gloria] was determined that she was going to go. She got out and found out different things, how to get there, and she went on her own. She was determined that she was going to go to school and she had that drive. She had that determination."

In 1970, Gloria moved to Houston, Texas and began attending Texas Southern University, one of the largest historically black universities in the U.S. The University was established in 1947 as a "separate but equal" state school for African Americans in Texas seeking professional education. It was there that Gloria Merriex began studying to be a teacher. After spending three years at Texas Southern, Gloria returned to Gainesville to earn her degree in teaching. She was an intern for a year after graduation and began teaching at Duval immediately. Gloria's two children, Tayana and Carl, still live in Texas, and Gloria has two granddaughters, 9-year old Carla and 4-year old A'niyah (in 2008) who, according to Jan, were very close to their grandmother. A'niyah's personality, in fact, often mirrors Gloria's personality. In these early years, "before she gave her life to Christ," recalls Cenia, "[Gloria] was a tough sister. She'd tell you off. She would tell me off and think nothing of it." Over time, Gloria seemed to rub off on her grandchildren, especially the youngest. "She was so into [her grandkids]," says Jan. "A'niyah would come visit me and I'd say I don't know what we're gonna do with this because she is a sista! She has [Gloria's] traits of being outspoken." Lilliemarie Harvey, who considers Gloria to be "like a Godmom," remembers hearing stories of this time in Gloria's life. "I didn't know her when she wasn't saved, but she talked to me about it all the time—about how she *used* to be...at certain times I would see a little piece [of old Gloria] come out and I'd be like oohh. She'd tell me that was nothing compared to [the way she used to be]...She always went head to head with everybody, she didn't care what anybody thought, but it was like 75 times worse [back then] - in a negative way. She was still a great teacher though."

Gloria also had many talents that developed throughout her childhood. "Jean loved to sing," recalls Cenia. "Coming up as children, her dad, myself, my oldest son, and Mary, all four of us would sing. [Gloria] never did sing, but now she sings in the choir at [Hope to the World Church]...after she got up and got into church then she sang with her praise team. Boy, she'd get loose on that praise team!" Leanetta (Lee) McNealy, the principal at Duval and a member of Hope to the World Church, remembers, "When she started singing at the church, that was a whole different side of Gloria. People just loved to hear her sing." Leon Marshall is a teacher at Duval who co-taught with Gloria. "Outside our meeting room there is a long bench and we would sing every day. She loved singing. That's pretty much how we would end our day. Sometimes parents and the kids and other faculty members would gather around to listen because it was just so wonderful."

Her talents certainly did not end there. "She used to sew her own clothes," said Lee, "[and] she made all of the graduation dresses for the girls [at Duval]." Cenia remembers Gloria staying up late to make use of her talent. "Oh yeah, she's got two machines in [her room], she would be in there up at night cuttin' out and sewing, making those dresses. She made those dresses...those corsages [the students] wore, she bought the flowers and put those together, she *made* that stuff."

"And the cooking," recalls Lee. "Those last couple of celebrations we had she did most of the catering for our staff...and those sweet potato pies!" Gloria's brother Charles remembers that Gloria "was a cook from the beginning." "Yes," says Cenia, "she was the one always cooking. She gets up and fixes breakfast. She knows [Charles] likes pork chops, so she would always cook. Sometimes she'd be fussin,' but she'd get up and cook."

Gloria also had some talents that were not as well known. "I remember when we...had a talent show where even the teachers had to get up and do something. [Gloria] must have been a drum majorette—you talk about twirling a baton! All up and between her legs and catching it behind her back...I'm thinking—amazing. Just amazing." This talent may have come from her years as a clarinet player in the Lincoln and Texas Southern bands. "She coordinated the dance routines when the band would go out and perform," remembers Cenia. "She was a good dancer...when she came back [to Gainesville], she was out at Eastside helping coordinate the dancers out there."

Perhaps it was these early experiences that shaped Gloria's beliefs about her students. Lilliemarie Harvey is in her fourth year teaching at Duval, and worked closely with Gloria during her first years at the school. "She knows exactly what [the kids] need and why. Just put Ms. Merriex as a child. She knows how she needs to be talked to, she knows what will motivate her, so she knows what to give [Duval] kids based on her background and how she grew up. So she's doing what teachers did with her back in the day."

Sometimes the kids had a hard time understanding the struggles that Gloria saw for them because of her background. "People think [our students] can't learn," says Lilliemarie. When they began acting like they didn't care in the middle of the year or if Gloria felt that students were not trying hard enough, "she would go to church with them. She would stop her instruction and just talk to the kids and say this is how people view you. This is why people view you like that. These are the steps you need to do to knock all those bad thoughts out. Don't take anything for granted, nothing's gonna be handed to you, you have to earn everything. She would stop for like 10 minutes and go to church with them, you know like the preacher part of it, and then move back to the math."

Leon Marshall agrees that it was her own personal experiences that shaped the ways in which she believed and acted. "She went to school with very little money, she had to work hard...so she understands struggle. So [the students] couldn't tell her anything about it because she lived that too...there were those who told her, oh you probably won't make it...that just made her more determined to tell you, 'oh yes I will!'" As such, Gloria would not let her students believe that they could not do what she did. In order for the kids to appreciate this, "they need to see—okay, this kid came from the projects...and they made 3 or better on FCAT, so I can do it also" rather than using their background as an excuse for failure. This mentality made believers out of children and, in turn, parents and the community.

"She taught them how to work together; how to always strive to be the best," recalls Lilliemarie. "That's one thing that a lot of students really don't know how to do—how to strive to bring out the best in themselves." Gloria wanted to show children that she came from the same place that they did, and despite the struggles she faced, she was highly successful. Lilliemarie remembers this as Gloria's message to the students - "[she] was a poor kid in a shack house who became this phenomenal, world-renowned teacher. Anything is possible as long as you're motivated and you stick to it, but nothing is ever handed to you, nothing is going to be given to you, you have to work for it and go after it, and that's what she did."

Upon her passing, Gloria was on the verge of national acclaim. In the spring of 2008, Gloria mentioned to me that a major foundation had recently approved a grant program that would spend a great deal of money installing cameras in Gloria's classroom so that her work could be streamed online. When Gloria told me about this project, she was unsure of her personal monetary benefit. "I think they just want other people around the country to be able to watch me," she said, "I don't know if they pay me, but I don't mind." Other projects, such as a curriculum based on her practices and a professional development module based on her strategies, were also in development.

Throughout her life, as is evidenced by her record of personal and professional successes, Gloria was invested in the things that were most important to her. These things, family, friends, school, and church, consumed all of her time, energy, and even money. "She jumps into everything and does it not 100, but like 3,000 percent no matter what it is," says Lilliemarie. Given the amount of time that the two spent together, Ms. Harvey was able to see this energy in every aspect of Gloria's life. "We went to church together, we worked together, we hung out together, we shopped together, and she was like that in every aspect of her life, it wasn't just teaching. She planned parties and functions, and she worked just as hard for her church as she worked in that classroom. There was no running out of gas or halfway doing something, it was 100 percent on everything. I have never, ever met anybody like that. Some people slip in certain aspects of their life, you know...but she was just 100 percent on *everything* she was involved in." As such, Gloria left quite a pair of shoes to fill—shoes which most people feel will never be filled.

Chapter 2

Understanding how they learn is one thing. Being a friend, being a momma, being a daddy, being an enemy, is another. You have to be the enemy at some time because you gotta get on them if they don't follow through on what they are supposed to be doing. I'm their friend because I am here to help them...[but] when they come in and be disrespectful I got to be their enemy . . . if they need me, if I can get it within my means, then I go out and get things for them...a lot of them need a little love, but they understand ***what I say goes.***

-Gloria Jean Merriex

Lee McNealy has been the principal at Duval since 1994, but has roots as an elementary school teacher in Gainesville. It was as a curriculum resource teacher at Duval, in fact, that she first met Gloria Merriex. At that time, Gloria was working as a first-grade teacher. "She just had so many ideas about how to work with children," Lee explains. "So many neat ways to bridge things over for the young children and struggling children...and boy did the children love Ms. Merriex." By all accounts Gloria loved the children too. It was during these early years that Lee believes that Gloria began formulating ideas and lessons that grew to draw so much attention more recently. "I just can imagine that she's always had these creative juices thriving in her... Gloria was the catalyst even back then," Lee says.

Though friendly, the two were not as close back then as they came to be. Over several years and after a stint as a teacher on special assignment (TSA) for the School Board of Alachua County, Lee made her way back to Duval after completing her principal internship in various schools in the county, and spending another year as an administrator at a different school in the district. It was when Lee came to Duval as principal that her life and Gloria's became so intertwined. Aside from their work at school, both were members of Hope to the World church, lived in the same community, and traveled all over the country as school leaders together.

When Lee initially returned to Duval, Gloria was working as a part of the 4th grade team of teachers, a group Lee coined "the Three Amigos" because of their close knit nature and varied styles of teaching. Though the Three Amigos worked well together, Lee remembers that things were not always so harmonious between the members of the team. "I can remember Norma[1] would always say 'we did it *this* way at [a school she had worked at previously].' Jean finally stopped her one day and said, 'Well, you aren't there anymore, you're at Duval!' That brunt way that Gloria could approach folk would just leave them taken aback, but Norma said after that encounter with Jean that she was right—she was no longer at the other elementary school. So Jean got her straight on that."

This type of confrontation and challenging of opinions was not foreign to Gloria. "She was just a straight forward person," Cenia Merriex recalls with a laugh. "She would tell you how she feels, and then she'd tell you she was through with it. She wasn't a two-faced person...she'd tell you off...ooh, she'd tell me! You always knew where she stands." Jan agrees. "She was outspoken and would say whatever was on her mind," she remembers, but Gloria would not seek out an argument. "If you wronged her, she would let you know, but she didn't go looking for a fight. She wasn't a bully." The Merriex women believed this was one of Gloria's defining characteristics and realized that she did not shy away from the truth, even when it came off as abrasive. "Old as we are, we can't handle the truth...and adults shunned away from her [because of that]."

This personality trait did not come and go as socially appropriate as it does with many people. When a participant in a conference session or a professional development seminar, for example, Gloria would not passively sit through a session if she felt the presenter was on the wrong track. At a mathematics conference to which the she and Lee traveled, "Gloria would say, 'Boy, I'm not learning a whole lot. I could show them a thing or two.' One session that she came out of, she said, 'I had to get that consultant straight.' [Lee] said, 'Well, Gloria, I hope you did it professionally.' She said, 'Oh, I did! I knew they were on the wrong track and I started telling them how it could be done differently and everyone started looking at me.' I just laughed." Lilliemarie remembers Gloria relaying several stories of this nature. "She was just very opinionated, and if she didn't agree with something she was going to let you know...she didn't believe in sitting back and being quiet—that's the only way people learn is if they're challenged, and she did it all the time."

Occasionally, Lee had to confront her about these outbursts. "Sometimes, I would have to pull her aside and say 'Gloria, that was not a good thing that you said.' And Gloria would say, 'Well I said it anyway! I said it!' She would say out loud what she *wasn't* going to do." During a particularly important year at Duval, a reading consultant, Sally[2], was called in by Lee. Though Sally told Lee that she was not able to help because of limited time, Lee talked her into coming to Duval on Saturdays to work with the faculty. Sally could have easily turned down the offer, but Lee pleaded her case, gaining Sally's support. During one of the first of the teachers' sessions with Sally, Gloria piped up, saying she did not need a teacher and that she should be released to go back to her classroom. When Sally told Lee about this incident, Lee pulled Gloria aside and said, "Look, lighten up, we need this lady. I'm sure you have some strong ideas, but if we just work together it will work." Gloria did not believe that Sally knew best at that point, but went along with the temporary structure as well as she could.

This perceived roughness translated into Gloria's teaching as well. She was a very tall woman, and her voice commanded the presence of a room. She took a very strict approach to discipline, having students without materials or with an attitude leave class and call their parents under her supervision. She did not tolerate any outside talking in class, though the children were so engaged in learning mathematics most of the time that there was little time or need for such behavior. She had high expectations for their mathematical ability and effort, and required that they perform at all times. Her reputation was well known throughout the school. Up and coming fifth-graders knew the expectations, and she let them know as she passed them in the hallway that they had better be ready.

Jan Merriex recalls the time that she put her own son (Gloria's nephew) into Gloria's class and how he experienced her toughness first-hand. He was in fourth-grade at the time, and Jan let him know that she was putting him "in Auntie Jean's class. He needed more forcefulness to be successful." Given Gloria's reputation, he was not agreeable to this decision. "He boo-hooed, saying 'Oh Mama, please don't send me [into her class]!' He begged me [but] I stood my ground and I'm glad I did. I put him in Jean's class, and you know that first day she starts with writing. Those writing skills. I remember looking through his binder and one day he wrote 'I'm in fourth grade writing class, the teacher is my Auntie, and it is a living hell.' Jean said, 'Oh! You're in a living hell, huh?' And he couldn't say nothing. He stayed with her in 5th grade too, then he was getting ready to go to [Howard] Bishop [Middle School] and you know what he says to me? He says, 'Mom, you think that Auntie Jean can go with me to 6th grade?' I said, 'Oh, wow.' He got to know her. That's what happens - people judge people just because they see them at that point, but she's just a teddy bear."

As Jan knows from this experience and others, underneath the tough exterior Gloria was very caring and nurturing. "I think that people...were always intimidated, but once they got to know Gloria they were in awe of her," remembers Lee. She always wanted what was best for the children, and that is what made her so great. "She may have approached folk with that rough edge, [but] within her she was a softie." In fact, she and Sally, the reading specialist who Gloria initially took issue with, became friends in the years after the initial incident. "To the end, they were the best of buddies," Lee recollects. "Even when Sally would come [to the school] and it was not for Jean's grade level, Jean would seek her out during the day that she was here, or have lunch with her, or bring in some materials for Sally to look at and approve." Jean also became close with her team, the Three Amigos, and built an intense bond with Lee despite their differences. "She and I have had our times that she has sometimes not agreed with me and I did not agree with her. But at the end of the day, we could always come back together and be on the same page. I respected her for that."

It was almost as if she were cautious about showing her truly generous nature so that she would not be taken advantage of. Many people, including her family, other teachers, students, even researchers from the University, depended on Gloria. Lee was very aware of all that Gloria did for others, especially relatives. "She just did so much for her family. It is such a loss to them because they depended on Gloria so much...for support and resources, her nieces and nephews, all of them...really depended on her." Gloria's mother agrees. "She carried this house—there were lots of things she did that people didn't know that she did—she carried this house." This generosity and sense of caring, however, was not at the forefront of Gloria's personality, even when she interacted with her family. Cenia remembers sitting on the couch as Jean was walking through the house. "I might ask her 'Jean, you got a dollar?' and she'd keep walking and just tell me 'I ain't your bank!' and keep on walking through the house. When she'd come back through on her way out, I'd still be sitting here, and she might hand me a five, a ten, a twenty, then keep walking and go on about her business without saying a word."

Though outwardly tough in the classroom, Gloria showed this same furtive and deep care for her students. Lee remembers that even though she might not show it outwardly, "[Jean] would do anything for the children—give them the shirt off her back; anything...she wanted to touch children's lives and make a difference. It went *beyond* the instructional practices because children who had no shoes or had no food or needed this, and she knew about it, then she provided that for that child. It mattered not what she did or didn't have, but she would give her last to the kids. When you form that kind of bond, and that kind of rapport with a child, they know. Children know. They can decipher. They know who really, really cares, and who really may not care. And they knew that Gloria Merriex cared." In return, she expected students to do as they were told, and show that they appreciated her efforts. "When she passed and we had the wake," recalls Cenia, "there were children just crying, saying [Gloria] was just like a mother to them...and a dad too."

Lilliemarie saw things similarly, and found that Gloria's strict rules of behavior and high expectations made the classroom almost familial. "You know how your family is hard on you because they care about you, but you know they care about you, you know they'd bend over backwards for you, and you know that they're only saying that to build you up. She didn't say, 'Oh, I'm only being mean to you right now so you can be stronger,' but the kids just felt it. Like, she'll get in their face and yell at them, then turn around and be like hey, baby. They knew that she cares and she's just doing it to better you."

Buffy Bondy, a professor of education and director of the School of Teaching and Learning at the University of Florida, has worked with the faculty at Duval for the past 9 years. She saw the same sense of caring in Gloria coupled with a seemingly harsh exterior. "If a person were to try to assess [Gloria] at first glance, they would be way off base," Buffy recalls. "They just would not have the real story. It's funny—in some ways she was an open book—she'd tell it like it is and let you know what she thinks about things - but in other ways there was this very warm and caring part of her that was expressed in ways that might be unfamiliar to some."

During her time at the school, Buffy was officially a professor-in-residence, essentially serving as a consultant and as an integral part of the leadership team. Her job there as she describes it, however, was not particularly well-defined, and she often did whatever was needed at the time. "It's hard to say what exactly I did [at Duval]. I felt like I was just a member of the staff there...a lot of what I've ended up doing there is just consulting with people on any number of things which can be classroom management issue, but it can also be 'I have got to talk to Dr. [Lee] McNealy about something. What do you think is the best way to approach her on this?' There was a lot of sort of strategy stuff that I ended up doing with people...I did a lot of problem definition and solving work, and that could be academic, classroom-based pedagogy sort of stuff, or professional development—I was always involved with that. My theme was always to not do too much. Just stay on message…hammer away at one or two things, but not do a million things at once…I was in and out of classrooms a good bit because teachers asked me to, I did a lot of consulting with the principal. I would sometimes meet with the leadership team just to be part of the thinking and effort…I saw my role as morale booster too. I would just drop in on new teachers and see how things were going." According to the faculty, Dr. Bondy's work did much to build the collaborative culture among teachers and leaders at Duval.

During her years at Duval, Buffy came to know Gloria's work well. "With Gloria I would try to just sit on the stool by the door and just watch for a while. She might want to show me something…sometimes I would write some things down that she was doing, but I would always check in with her." Over time, Buffy wrote several papers focusing on various aspects of Gloria's instruction; however, her initial impression of Gloria's instruction was mixed. "When I first was at [Duval] and I'd sort of peek in there now and then, I have to say that I felt that she was maybe a little bit scary. She was a really tall woman…[and] from a sort of white, middle class perspective, not a warm and fuzzy someone who reaches out, though really very warm and very funny and very fun to be with. But, you might not see that at first…I could see people thinking, including some student teachers, that she was intimidating. Then, if you really pay attention, you realize that you need to be watching how kids respond to that because your response really doesn't matter so much."

As pre-service teachers and teacher educators observe others' instruction, judgments are often made about what goes on in classrooms. Many times, this leads observers to draw conclusions about people, particularly teachers, which exist and work outside of our own cultural norms. Gloria, who was easily misunderstood and frequently described as intimidating and harsh, was so in tune with the cultures and realities of her students' lives, that she knew how to communicate expectations effectively and consistently. Because of her own background, she knew what they needed in a teacher and mentor, and refused to conform. "I'm not going to beg them [to do the work]," Gloria would say. "And [I] understand where they're coming from and their home life. I'm not gonna teach them to be white—that's the bottom line."

Moreover, Gloria understood the struggles of the families in the neighborhood. Having lived through such perceptions of inadequacy herself, she knew that many of her students had been told in some form or another that they could not learn, thus hardening them to academic contexts. "Gloria knew that you have to understand the culture in order to formulate your lessons," Lee recalls. "She had lived some of the things that some of these children lived…so even though she was the teacher, she had not forgotten from where she came. She could speak their language [meaning] she could interface with the family, but did not demean them in any way. She had established that kind of trust." Moreover, Gloria made sure to keep in close contact with members of the community, making the school central to the neighborhood. Lee remembers her first days as the principal at Duval. "Gloria got in my car and we rode to Duval Heights...she said, 'Well, we're gonna go and meet the families.' She took me under her wing and, as we went from house to house, she was introducing me to the families and saying this is the new principal...she knew the families, they knew Gloria...they respected her. The connection was far beyond the school...she knew where these kids were coming from."

This connection allowed her to break through students' tough exteriors and tap into their potential using a demanding style that could be perceived as harshness. Outside perceptions, however, are not necessarily important to student achievement. "What you see," Buffy continues, "is that the kids really respect her and really want to live up to her expectations. If she reams them out over something, they may look down or a couple of those boys may grumble, but more often what you see is they buckle down and they get on it...what we may see as reaming out or harshness, [the kids] don't interpret in the same way. What they see is a teacher who really cares about them who is going to make sure they are successful. She communicates how committed she is to them and how much she cares about them from the beginning, so then, when she has to get tough, which she will not hesitate to do, they respond to that because they know that this is someone who has their very best interests at heart and really wants to see them excel. And they start thinking of themselves as people who can excel."

Given the tragedies in her life, it is not surprising that Gloria expected excellence from her students no matter what their situation. As someone who came from the same neighborhood and experienced mandated segregation, she would not accept excuses of any kind. "You can't play with them," Gloria once told me, "because then when you need to get serious they don't understand." One of the most notable losses in Gloria's life came with the death of her father. "It really hurt her when her daddy passed," recalls Cenia. Lee recalls that it was difficult to get Gloria, who was typically consumed with her teaching, to come back to work after this trauma.

Gloria experienced several other formative misfortunes in her life. After finding a way to get to college, she was faced with the challenge of staying the course when one of her good friends, Ricky, passed in 1972 during a visit to Gainesville. As Cenia tells it, "[Ricky] came down here to visit with her and...we went down to a park there in Keystone heights. It was on a Thursday, and it wasn't five minutes until that boy got drowned there. So we had to call his folks—he was like family. [Gloria] didn't hardly want to go back to school, but she went." In the years following the tragedy, Gloria met another young man named Ricky, a basketball player from Alabama. The two grew close and remained friends after graduation. "When he finished school he was playing ball...overseas. He kinda got tied up with something, but he got killed also. I think that kinda knocked [Gloria]."

These types of experiences may have hardened Gloria on the outside, making her appear to be confrontational and emotionally bulletproof. "I think she built up [her tough exterior] as a child because she had to fight her way through," recalls Angie Terrell, the fine arts coordinator at Duval and a close friend of Gloria's. "She was always the underdog, and that taught her to have tough skin." Moreover, as she began to gain notoriety later in her career, she would often speak about people from various places that might be taking advantage of her generous nature and willingness to share. This may have shaped her interactions with new observers, making her seem hardened on the outside. At the end of the day, however, she was happy to be helping children.

This carried over into one of Gloria's favorite hobbies—shopping. "Her first love was to the kids, learning, and math. Then shopping! I don't believe a day passed that she didn't buy something. She may think about it and take it back, but she still bought it," recalls Jan. "They knew her in all those stores...we'd go to Jacksonville, Tampa, Orlando, she'd be everywhere." Lee McNealy remembers, "You could find Gloria either at the mall or Duval. Those were her two things." She and Jan would shop for hours, going to Jacksonville, Tampa, and Orlando. "She'd be everywhere. I just couldn't hang with her. Every now and then I'd gear up, take my pills, and take something to keep me strong all day long because when we'd shop from...if we'd leave here at 8:00 in the morning going to Tampa, Jacksonville, something like that, we don't come back until, what, about 11:00 at night. We'd close the malls. Especially during the holidays. She'd shop for [her grandkids]. Her children are adults and she'd still shop for them."

As a result, people at various stores knew Gloria and would contact her for special events and sales that might be happening. In return, she gave these stores her business. If she did not feel that she was treated properly by a store employee, she would make her dissatisfaction known. "If they didn't talk to her like she felt she should be talked to, oh she'd write a letter on them in a heartbeat. She always got an answer too," recalls Cenia. "If she didn't like the way she was treated at the store, she wrote," says Jan. One man, a store owner, took notice. "When [Jean] walked in that store, they knew uh-uh, don't even mess with her. They'd just say, 'Hi Ms. Merriex!'—they knew."

Gloria would not make her personal problems public. Her mother recalls that there were things that even she did not know about Gloria. "She was a private person about her life and...her problem was her problem and her business was her business," says her mother. This privacy seemed to be due in part to Gloria's selfless nature. Jan is still learning about things that Gloria did for others. "I met a man [after Gloria's death] and he was telling me how his wife was real sick and [Gloria] would come to the house and do what she could do for his wife. We didn't even know that." It seems that was just the way she was. "If she knew you and she found out you were sick, this is something she actually did. And we'd find out because we'd see somebody and they'd say 'oh Jean came by and she cooked me something' or 'she cleaned up' or 'she did this' or 'she did that,' and I'm like really? Yeah, that sounds like her." Cenia recalls that Gloria "was the person that when she did something for you she didn't go around and broadcast it. There were people out there that she did things for and we didn't know nothing about it. Why? Because she felt like it wasn't your business. That's just the way she was."

This generosity spread from family to community. "We had an aunt—she would always say 'Jean *always* comes to visit me and y'all never come!' Even in the Duval community [Gloria] would come to the house [of a family in need]. She didn't just teach; she was caring even more than we would ever really know," says Jan. If she found out something had happened to someone or their family, she made an effort to contact that person. "She would call and say 'how are you doing today?' Lots of people don't understand sometimes that little things are what count. It's not the big things, it's just the words, you know," recalls Cenia.

Gloria's private demeanor was not always a good thing, however, as those close to her learned more recently. Jan thinks this sense of privacy was "good in a way but it's not good in other ways. If she had something going on she wouldn't tell us—she wouldn't say anything." This was especially relevant when it came to Gloria's health. "She was on insulin for diabetes," says Cenia. "She didn't like that. I'd say 'Jean have you checked your sugar?' and she wouldn't say nothing . . . [because] she was used to being independent. She didn't like being dependent at all...you didn't ask her about her personal business, now. Cause she will tell you off and keep walkin'!"

Lee also worked to encourage Gloria to pay more attention to her health. During one particularly jarring episode after which Gloria ended up in the hospital because of the high glucose level of her blood, Lee called her own doctor and begged that she take Gloria on as a patient. In the short term, Gloria got much better. Over time, however, she stopped going to the doctor. "Gloria did not take care of herself...she brought a letter in from our physician at the time and said, 'Read this, Bosslady.' The doctor was saying that she was releasing [Gloria] from her care because she could not be liable. Gloria was not following through and was not keeping her appointments. That was a major blow for me." This type of pattern was common for Gloria. She would be feeling bad, be forced to go to the doctor's office, get better for a while, and then stop taking care of herself. Leon saw this as a side effect of her undying commitment to the school. "She was the glue, I would say, that made everything work, and she really denied herself so many times—health-wise, everything, for the community, the kids, and the school."

Getting Gloria to the doctor was another challenge. "Even when she was so ill a couple of times in dealing with her diabetes [her family] couldn't get her to go see a physician. I would go over there and just tell her mother or her brothers, 'give me her pocketbook, put it in my car—she's going.' And she would always laugh because I would say, I *need* you at school. And so you are *not* going to be sick," recalls Lee.

The only other person that could get her to move was her childhood neighbor and fellow educator John Dukes. "When she got sick, a lot of times they would call Mr. Dukes to come and make her go to the doctor," recalls Cenia. "That's the only way she would go. Lots of times when he'd get here, she'd say, 'Where you goin? Who called you?' He'd say, 'Let's go.'"

Jan recalls that this was a reciprocal relationship between Gloria and Mr.

Dukes. "I remember one time when he was sick, we went over there and he was home alone. He said, 'I'll just wait to go to the doctor,' and [Gloria] said, 'No you're not! You're getting yourself up right now and we are going to the hospital right now.' He didn't exchange words with her, 'cause he knew. He'd put on his little shoes and kinda look up at her—like a little kid! His daughter lived in Atlanta and she would call the house and say, 'I need you to go over there and check on Dad 'cause he didn't sound good.' I would say, 'Let me go get Jean because he's not gonna move for me. I know if I bring her, he's gonna move.'" "That's right," recalls Cenia. "When he was sick she'd go sit by him, and when she was sick they would call him and...see they didn't ever call me, they'd call him and he'd go out there and get her."

Thus, despite her seemingly tough exterior, Gloria was a very caring individual who cared deeply for those around her. There has been speculation in the community that because of her attention to others' needs, she may not have taken care of herself appropriately. Leon recalls things differently. "She said, 'When it's time for me to go, the Lord will take me.' She didn't want anyone making a fuss over her, and I think she knew that day [that she passed away] that it was her time."

Chapter 3

I've been teaching so long, I don't even know [what influences me] anymore. I guess just getting' up every morning, coming to work, and ***making a difference****...then seeing the influence I have on the kids and being proud of them once they finish and they understand. It makes you feel good - especially if you have your heart in it. If you don't have your heart in it and you just want a paycheck it doesn't matter. You know I'm not in it for the money!*

-Gloria Jean Merriex

When speaking to the faculty and staff at Duval Elementary, an overwhelming camaraderie is evident. This is largely due to a committed and inspirational leadership team coupled with the drive of each individual teacher to work hard in educating this particular group of children. The school which is thriving now has not been without its struggles over the years.

The No Child Left Behind Act (NCLB)[1] of 2001 introduced new accountability and grading systems in public schools. These school assessments are meant to give the government a benchmark by which to measure schools' successes in educating students, and allow for comparisons between institutions and between students. Additionally, Florida's A+ plan mandates that schools are scored on an alphabetical scale. Student test scores, based on tests that have been developed by state systems in accordance with NCLB, are the data from which these school scores are derived. These scores are used when publically evaluating schools, students, and teachers. As such, these assessments greatly influence student retention, and inadequate performance can even keep a student from graduating high school. Further, the school as a whole is given a grade (A is the top score while F is the bottom score) based on such results. This not only affects the public view of a particular school, but affects the government's role in mediating school activities such as curriculum, instruction, funding, and special programs. Because of the large amount of weight given to the scores that result from these assessments, they are often referred to as high-stakes tests[2].

In Florida, school grades are largely based on student performance on the Florida Comprehensive Assessment Test (FCAT), and also take into account adequate yearly progress (AYP). Essentially, the state looks for passing scores on the FCAT (a score of 3 out of a possible 5 is considered passing), and gains on the test particularly in the areas of mathematics and reading. Learning gains are made, for example, when a student who scores a 1 (out of 5) one year improves to a 2 (out of 5) the following year. The state also specially considers students who fall in the lowest 25 percent in terms of scores. Schools that do not make AYP are required to develop and submit a School Improvement Plan that outlines the ways in which these needs will be addressed, and particular students receive an Individualize Education Plan (IEP).

NCLB also includes many sub-sections and specific plans for action. As outlined in the official document, title I is meant to "[improve] the academic achievement of the disadvantaged."[3] Duval qualifies as a title I school, meaning that there are a high percentage of students in the school that qualify for the free or reduced lunch program. Essentially, the school serves, as the government calls them, "economically disadvantaged" students and families. In an effort to "[ensure] that all children have a fair, equal, and significant opportunity to obtain a quality education and reach, at a minimum, proficiency on challenging state academic achievement standards and state academic assessments," the government allocates money to these schools, which serve the "needs of low-achieving children in our Nation's highest-poverty schools." Historically, schools with this population have struggled to succeed, and often economically disadvantaged students do not receive the same quality of education as the economically advantaged. It has also been argued that the accountability grading system is unfair to these and other populations (students who do not speak English as their first language, for example), as the FCAT is written without regard for cultural or linguistic diversity (Jones, Jones, & Hargrove 2003).

During the 2001-2002 school year, Duval earned a grade of F under the state rating system. Lee recalls this time with emotion. "The day that we got the F [Gloria] came looking for me…she kept asking what was wrong, but I just couldn't voice it." After what had been a year of hard work and determination, this came as a blow to the school. "I just walked away [from Gloria] and she followed me down the hall…she was not a touchy person, even though she was caring…she just grabbed me and hugged me and said, 'What's wrong with you?' I said, 'Gloria, we made an *F*.'" The media portrayals were harsh and painful to the faculty, and brought the morale of the school down. The school also faced the possibility of losing a large grant that they had just received as support for integrating the arts into the curriculum. After their initial hurt and devastation, however, Lee and Gloria got mad. "We've had our tears," Lee said, "and we've sobbed, and now we've got to get busy…she was as determined as I was and we had to get everybody else to be determined… we were so determined to show everybody—the district, the state, our parents, the community—that we were not failures."

The school immediately held a town hall meeting in the auditorium for parents. Gloria spoke and Lee spoke, assuring parents that "The F will not determine who we are. Just have faith in us." The following school year an incredible turn around occurred, led by Lee, Buffy, many professional development facilitators, and the entire faculty at Duval. Gloria was also a large part of that change, recalls Lee. "That year we just worked so hard, and she worked the hardest, I know she did. That's when we started seeing those gains, and she worked night and day. When we moved from [a school grade of] F to A, and schools started coming wanting to see what we were all about here…some of our sister schools would come over, and we would always include Ms. Merriex in the tour…and they would always be amazed. When we would have our debriefing in the afternoon, questions would be asked, but it would always gravitate back to Ms. Merriex's classroom."

Fortunately, the school was able to keep the grant that they had received to infuse the arts into the curriculum. Gloria, with her rhythmic teaching, chants, raps, and movement activities, was ahead of the curve on this front. As part of the grant, a team of representatives from the school was required go to a summer institute in Mississippi. These institutes were meant to help schools to find ways that the arts could be incorporated and woven through all subjects. Gloria and Lee traveled to these institutes together. Focusing on the arts "was like putting Gloria in the Briar Patch, because this was what she knew anyway. The Mississippi commission and the teachers and educators, they were just budding with this, and we had already been taking that in her practice, so she just ran with it, and taught them one or two things when we were in Mississippi," recalls Lee. "It was just amazing." Because of her expertise, many of her colleagues feel that she was under-utilized by the district. "That is one of the saddest pieces," recalls Lee. "It was awful because they would pay other big time consultants to come in, but because she was employed by the school board they would not pay her [to consult with teachers]...so even the people here in Alachua county didn't know what a gift we had in our midst."

Gloria's success was central to Duval's achievement in mathematics. "She was the driving source. She was the vehicle that put Duval on the map. The community, the city of Gainesville, Alachua County, the state of Florida just had high admiration for her because she just spoke loudly in the educational arena with her practices," remembers Angie. The new state assessment system gave Gloria a new goal: to help her students succeed in the eyes of the state without having to abandon their culture. Gloria worked tirelessly to this end, constantly thinking of innovative strategies that would help to improve her students' mathematics scores. Each successive year, her students showed the highest mathematics gains in the state, their overall scores at the top. In fact, Lee McNealy recalls how Jean thrived on working with the most challenged children. "That was her challenge," she says. "[Jean] glowed with the children who were, I would say, the rough piece of coal and then became the diamond after she completed her work with them." Leon Marshall found the same thing to be true. "She loved the challenge of kids who were [labeled] ESE[4]," he remembers, "who are supposedly on a different pace. She loved the challenge of showing every child can learn. [Teachers] just have to be flexible, we have to be open in finding ways to teach them, but every child can learn."

Buffy remembers a similar dedication. "She was really interested in kids who were deemed to have learning disabilities or kids who were supposed to be receiving special education services. It was kind of like she had something to prove there and they had something to prove as well. So that whole thing, this belief in individual children and in a people to excel, because it wasn't just to achieve, it was really a belief in excellence which I think is an important point. It wasn't just 'they can learn,' it was 'they can really excel.'"

Lilliemarie remembers this same unwavering standard of excellence despite the varying levels of students that Gloria had in her classroom. "Her standards for the kids were through the roof. Most teacher's standards are like 'oh, maybe you can do it, if you don't we'll adjust.' There was no adjusting for [Gloria]. Her standards were way up here and you were either going to meet it or exceed it and that was it. That's just how she taught her class and how she lived her life with these kids." This unwillingness to lower the bar was almost always met with success. "Children become exactly what their teacher perceives them to be," recalls Angie Terrell. "If you've got this laid back teacher who just goes with the flow, eventually that's how the class gets. If you've got a teacher who comes in the room with a whole bunch of issues and doesn't really know how to work with [the kids], then she's got a room full of problems. [With Gloria], you've got this teacher who raises the bar and says, 'I have a class full of gifted students,' and midway through, all of the students are acting as though they're gifted."

Gloria often disagreed with diagnoses given to children such as "learning disabled," and worked tirelessly to help such students achieve excellence. "He wasn't learning disabled in my class!" Gloria would say when asked about particular students who had struggled throughout their schooling years, implying that they were mislabeled. She would always say that it is often the teaching or teacher, not the learner, which is the source of any academic or behavioral problem. Gloria would spend hours observing her students and determining how they took in information, always asking "What makes the kids work? What makes them tick? What makes them think?" These questions drove her mathematics lessons, her constant revisions, and her decisions in the classroom. "I don't care what level an ESE kid is," Leon remembers Gloria saying, "they can learn the lyrics to a song and they can sing them, and they can remember a beat, so they can learn math in that way as well."

Angie Terrell recalls that "[Gloria] could see a brightness in a kid that so many professional people would overlook. You know her Phillip[5] story—he was a little boy that was an ESE student and right away when he came to her she knew that the kid was gifted. She spent many, many hours making sure that kid got out of the ESE program, and now he is at Eastside in the IB[6] program.

Jan, who works with handicapped children, remembers that she and Gloria would often swap ideas about how to work with kids who seemed to learn slower than the others. "She didn't shove them to the side; she made sure that they got [the mathematics concepts]. She even put them on her Math Team...she would build that confidence to say okay, you *can* do this. I know it looks hard now but you know you *can* do it. You keep striving, and if you fail, push it back." Just as Gloria would never tell Jan why they could not use the front door at Videll's drugstore, Gloria would never tell a child that they were excused because of a label; rather, she pushed them to excel and accepted no excuses.

FCAT testing gave her even more motivation to focus on these students, as she could see on the test itself that the child might need something more than had previously been provided. Her belief that it was the teaching, not the child's ability, drove this focus even more. "The children that some people would put aside and say they can't learn, she was determined to show that they could learn if given the right instructional practices, the right nurturing. She would not give up on any student," recalls Lee. The results of this drive were always evidenced when FCAT scores were released. "Her scores were just phenomenal. This year [2007-2008], 80 % of the kids passed with a 3 or above—she was the highest in the district, probably in the state," says Mr. Marshall. "I thought, what a beautiful tribute to her to exit out on. She left without choice, but she left a legacy that we can look back on and always remember her."

This attitude of persistence was likely instilled in Gloria many years before she was known for her innovative mathematics teaching. In the early years of Gloria's teaching career, the Title I schools had a pull out model. This meant that throughout the week, the kids who were identified as eligible for Title I services would come into a separate lab classroom for extra support. Initially, Gloria worked to help children who struggled, working in the Title I lab.

Gloria also worked with students from other grade levels, teaching fifth grade mathematics during the year, but also continuing to teach reading in the mornings and younger children during summer school. Her success was impressive with each age level. "It mattered not the age level or the grade level. She could take these children and make them feel so good about themselves and how smart they were," Lee recalls. This empowerment and self-confidence building through culturally relevant practices was a hallmark of Gloria's instruction. Children would come to her having been "expected to fail by society," as Gloria used to say. This determination and belief in children, no matter what their situation, motivated Gloria's teaching methods, and drove her to continue on in pursuit of excellence.

Gloria also wanted to help sustain the community surrounding the school. Because of the city's history, students on the East side of town were statistically more likely to have fewer opportunitics, possibly dropping out of school. "If you think about a kid who can drop out of school at the age of 16," says Leon, "if you don't put some kind of foundation by the time they're in 5th grade, they've got [only] two more years and they're gone. So we'll have a lost generation and a dying community because nothing is gonna come back to the community—it's gonna remain the same. Just like a farmer...if we don't sew some seeds in this community, it's gonna die out, and then it won't be a community, then there won't be any schools in the area." In other words, Gloria saw teaching and producing successful learners as a harvest. "Did every seed come up? No. But enough seeds came up so that we could be an A school, and that Lincoln and Howard Bishop could be A schools."

This focus on struggling students speaks to the theme of equity that resonated throughout Gloria's practice. She did not want any student to go without. Outwardly, this meant that she would do everything she could to provide for children who needed food, clothing, or money. Jan remembers that Gloria would search high and low, even calling on her family, to find appropriate clothing for Duval students. "She would call me and say 'I need a pair of black shoes—whatcha got?' She called every house she could think of trying to get what those kids needed."

Leon remembers the same type of generosity in the school setting. "Ms.

Merriex was first a resource for the students. If a child needed a shirt, pants, money, food, tutoring, she was there for them. Her pocket was just like a recycle bin. It was always going. So nobody could say I don't have pencil, I don't have paper, I don't have decent clothing, because she was going to make that her business to make sure. And she did it looking for nothing in return. The journals, she went out and bought every child a journal at the beginning of the school year. She bought the journal, she bought the cover for that journal. It meant that much to her to know that every child, when they came into her classroom, they had the same resources."

One amazing thing about Gloria's ability to instill a drive for excellence in her students was that she did not directly address issues of societal perceptions based on race and income directly. Though she might occasionally mention something to this end, such comments were few and far between. In other words, she did not constantly hit the children over the head with the idea that they had something to prove (though she occasionally made direct statements). Rather, she let her actions speak for her. Further, when she did help children acquire material things, it was done respectfully. "She knew if a kid had a pair of shoes on that were too small," recalls Angie. "She wouldn't broadcast anything, but in a subtle way she'll say, 'go out to my car and bring me a bag' and that kid would go home with a pair of shoes that fit. It was no big deal…it was done in such a way that would be appreciated and not belittle a person or anything like that. So those are the kinds of things that she really understood."

Over time, teachers from the district began coming to Duval to watch Gloria in action. Soon, University faculty, doctoral students, preservice teachers, and grant organizations from across the U.S. followed suit. Several groups began videotaping her lessons for various reasons, and using them for a multitude reasons such as preservice instruction, professional conferences, and inservice workshops. It was exciting that these sorts of projects were finally coming to fruition because "she wanted to be recognized for the work that she did," recalls Buffy, "[which was] kind of an interesting thing as a woman in this society where we often are underselling ourselves, she wasn't underselling herself." At the time of her death, her notoriety was just beginning to reach another level, and several projects were set up and funded so that the sharing of her ideas would be more beneficial to everyone involved, especially Gloria. Lee believed that "in years to come, if Gloria would have had the opportunity, she probably would have started her own school too. I just really think that."

Leon summed it up nicely, saying "every time we had a dinner or a gathering, she was gonna do the chicken wings, she was gonna do the rice, she was gonna do the green beans, she was gonna do the seafood salad. She made the pound cakes, she made the potato pies. Why? Because she thought she was a good cook, and she was, but she wanted to make sure there was enough. I don't care what you bought, she was gonna make sure that no one came and didn't eat. She was that type of person." Similarly, Gloria wanted recognition for her teaching because she knew she was special and gifted and wanted to show her skills, but also because she wanted no child to go without self-confidence and excellent, relevant mathematics instruction.

Gloria often stated to me that she was not sure of the sources of her motivation. What she did say was that she was influenced and driven in large part by her surroundings, namely her family, the children, the community, the school, and the church. Her passion seemed almost innate, and she believed that she was meant to teach children how to excel in academics and life. "They say I was made for teaching," she once said in an interview, and the satisfaction that came with fulfilling this role was enough for her. Lee recalls, "That was her goal—to teach. And she didn't want to do anything else! You know, she could have, but she wanted to teach. And she wanted to touch children's lives and make a difference," and that is what drove her throughout her career.

Though she rarely spoke about outside sources of inspiration, Gloria was quite taken with well known educator Martha Crosby[7]; she is best known for her work and success with impoverished children, particularly in inner city Chicago. Gloria read about Ms. Collins' methods and ideas, and aspired to share Martha's mission to educate inner city children. "She really studied...Ms. Collins, her practices and what she had been able to do with the children in her school," recalls Lee. The pair once took a trip to a nearby county to see Ms. Collins speak. "We were *all* mesmerized with Martha" and the great success she had in opening Westside Preparatory School, which was founded by Ms. Collins in 1975.

Westside Preparatory School is located in a disadvantaged Chicago neighborhood called Garfield Park. The opportunity for Gloria and Lee to visit the school arose while the two were in Chicago for a conference. They broke away from the rest of the group, and "They let us observe all day," Lee remembers, "and it was just amazing." When they later compared notes from the day of observation, both women were flooded with ideas, but sat back and realized that Gloria was already doing a lot of the same types of things in her own classroom. "It reconfirmed for both of us that she was on the right track."

Though Gloria and Martha Crosby shared similar aspirations and accomplishments, the two varied in style. Their work, however, seems to strive for the same end—to encourage reasoning and thinking among traditionally underserved students. Ms. Crosby utilizes the Socratic Method, using questioning techniques and abstract ideas to encourage thought and discussion. Gloria also introduced abstract concepts of mathematics, but used familiar cultural tools and practices to help children access the information. Through this and with much repetition, she was able to encourage reasoning, questioning, and confidence in mastery. When she taught reading, Gloria came up with the "prove it" method, through which students were made to read a passage, answer questions, and prove their answers to the class.

The two were similar also in their beliefs about children in high-poverty neighborhoods who had often received sub-par teaching. Martha once wrote, "I have discovered very few learning disabled children in my three decades of teaching. I have, however, discovered many victims of teaching inabilities." Similarly, Gloria felt very responsible for her students' learning and believed that all teachers should be accountable, though many shunned this responsibility and blamed children for the lack of learning that occurred in the classroom. "[Some teachers] don't understand that the lesson is not for them," Gloria said. "Once you understand that you've got to move away from your old-timey teaching and see how kids today learn, then you'll be successful. As long as you stay locked in to where you taught many years ago, you'll never be successful...at some point in time you've got to satisfy the students and you gotta satisfy yourself with the lesson." This unwavering belief that each student is able to achieve was fundamental to her practice.

John Dukes was another source of inspiration for Gloria. When Angie Terrell was in high school, Mr. Dukes was her class sponsor. She remembers Gloria speaking of him with high regard. "He had a great impact on her life...His family were their next door neighbors and his family took on her family like his own. A lot of the things that her family couldn't afford, he provided for them. She just had such high esteem for him and I think he's one of the reasons she became a teacher because he was such a great mentor for her." It seems that Mr. Dukes' skills as a math teacher also had an impact on Gloria. "[Mr. Dukes] was a dynamic math teacher and he would make all the kids at Lincoln High School excel. It wasn't were you able to do trig or geometry, he decided that you were gonna do it...and we did it!"

"She loved Mr. Dukes so," recalls Lee. I remember the same sentiment. I will never forget the first time that she told me about Mr. Dukes during my first year of working with her. She mentioned that he was a neighbor of hers who was a teacher when she was a young girl. She and he grew quite close over the years, and she would often visit his home. "I would go over there and I noticed it was cooler. I realized that he had air conditioning. We didn't have air conditioning!" As a result, she made the connection that as a teacher, you could afford nice things like air conditioning, and she made it her goal to someday have that luxury herself. "She always said that he inspired her," recalls Jan Merriex. "She found him over at Duval and had been there ever since." John Dukes eventually became an assistant superintendant in Alachua County, but his influence at Duval Elementary is not forgotten. Several years ago, a multipurpose building was named in his honor at the school, and this is where many of Gloria's events for the children would be held in recent years.

Part II: The Classroom—A Research Perspective

The notion of "cultural relevance" moves beyond language to include other aspects of student and school culture. Thus culturally relevant teaching uses student culture in order to maintain it and to transcend the negative effects of the dominant culture...the primary aim of culturally relevant teaching is to assist in the development of a "relevant black personality" that allows African American students to choose academic excellence yet still identify with African and African American culture.

-Gloria Ladson-Billings,
The Dreamkeepers, 1994, p.17

Chapter 4

> *I just sit and watch [the students] sometimes. I notice that kids can rattle off a song, even a new song, just rattle it off. I see it happen and think my God how can these kids learn a song so quick? Because they hear it all the time. They hear it every day. So I noticed this I said* ***my teaching style is gonna be like that****. So I teach [these concepts] every day. And even if I teach something new, I go back and reintroduce the old, and combine the old and the new together. So now they understand when they walk in they have to be straight, when you're straight you'll understand the processes as well as how to how to solve the problems.*
>
> -Gloria Jean Merriex

During my years as a graduate student I taught a mathematics methods course in the teacher preparation program. Essentially, this course was meant to teach teachers how to prepare and enact effective and engaging mathematics lessons in an elementary school setting. Pre-service teachers take this course during the first semester of their fourth year in the program. In the fall of 2006 I taught the course for the first time, and the population of my class was representative of most teacher education courses at the university: 28 white females, 2 African American females, and 1 Latina female. In subsequent semesters I taught the course three more times, and each time the percentages were similar. Several semesters I had one or two white males in place of one or two white females, but the overall population stayed quite constant.

On the first day of the course, I always ask my students to respond to the following prompt: "Describe your ideal classroom. What are you doing? What are students doing? What is the overall environment?" Nearly every student responds the same way. These ideal classrooms are described as engaging, supportive, and student-centered with a great variety of instructional techniques (e.g. group work, direct instruction). Students are engaged in whatever it is that they are supposed to be doing, and an appropriate level of noise is tolerated. If students are doing group work, the teacher walks around to help students, asking guiding questions and supporting interaction.

Phrases such as, "How is your group doing Johnny? Do you have enough materials? Are you sure that is the way you want to go with this problem?" are common and ideal. In terms of discipline, my students are careful to be sensitive, saying that they would ideally take a soft tone with students asking, "Are you sure you want to throw that crayon? What are the consequences going to be if you do that? Can you please go pick up that crayon?" for example.

My students have the best of intentions when describing these classrooms—these are the ways in which they have been schooled, and the ways of communicating that they describe are consistent with their frame of reference. In fact, to most current teachers this may sound like the ideal environment for learning.

They, and my students, think of this type of classroom as normal, and argue that they are just trying to be fair to all children. What is not as evident is that such classrooms are highly cultural and are based in the comfort zone of the teacher. In other words, white culture drives the norms enacted in "ideal" classrooms. White children might respond well to softer directives, for example, simply because such modes of communication mirror their culture at home. Such techniques, however, are not consistent with the home culture of many African American children who are used to a more directive tone. This type of cultural incongruity may confuse an African American child, leading him to resist and act out.

Teachers are quick to claim that they have set up a fair, caring, and equitable classroom and that the African American children are in the wrong. In some cases, teachers will even go as far as to make an unfounded generality about the value of education in the eyes of the African American community. Such beliefs only exacerbate the dominance of white culture in our schools and highlight the need to explore successful teaching practices in diverse classrooms.

Culturally Responsive Teaching (CRT)

The teaching methods of Gloria are extremely complex and difficult to describe with words alone, but do provide insight into culturally-based teaching methods. The transformative nature of her instructional practice, however, makes it necessary to use accepted research methodologies to deconstruct her instructional practice in ways that are meaningful and relevant to schools, teachers, and administrators. Successful, African-centered pedagogies have been studied (Ladson-Billings 1994; Irvine 2003) in other contexts, largely under the umbrella term of culturally responsive teaching (CRT[1]; sometimes referred to as culturally relevant teaching or culturally sensitive pedagogy). Culturally responsive teaching practices in largely African American contexts "[seek] to recapture control of the education of African American communities" (West-Olatunji, Baker, & Brooks 2006), using culturally-based communication patterns, body language, and strategies to empower students. CRT provides a useful framework through which critical examinations of current ideas about culturally responsive pedagogy can be refined and expanded to include mathematical contexts. In order to give context to Gloria's work and add to the existing literature, we will first take a look back to the general framework of culturally responsive pedagogies and classrooms, largely in relation to African American learners.

CRT is discussed within the American historical context, and gives a framework to much of the work of multicultural educators. Culturally responsive teachers view the incorporation of cultural perspectives as a necessary educational commitment. Students are viewed as cultural beings with cultural filters (Gay 2000) through which all information is sifted. In other words, cultural experiences and identities are seen as the foundations for all other experiences and behaviors (Ladson-Billings 2001).

Educators who embrace CRT aim to capitalize on the fact that varying cultural filters exist among students and teachers. Moreover, these educators aim to resolve, through empowerment and validation, the dissonance this variation causes. As mentioned above, individual culture and experiences are the basis for all other actions (Gay 2000; Ladson-Billings 2001). Often, school cultures (and teacher cultures) are not consistent with student cultures, causing student achievement to suffer. This implies that inconsistencies in school versus home culture (e.g. how learning is assessed, how information is acquired) is as much a factor in students' academic struggles as lack of ability (Boykin 1986; Spindler 1987). Moreover, this indicates that poor "test scores and grades [among students of color] are symptoms, not causes, of achievement problems" (Gay 2000, 16).

Culturally responsive teachers recognize cultural incongruities as the cause of the achievement gaps, and can therefore intervene appropriately. Too many current pedagogical solutions in schools focus on standardized testing as the root of the problem and result in prescribed curricula focused around these assessments. When we consider that cultural discrepancies are at the root of the problem, it becomes clear these types of dictated pedagogy and curricula may actually exacerbate the achievement problems of students of color.

Teachers who teach from a culturally responsive framework recognize the strength in each student's funds of knowledge, and "is very explicit about respecting the cultures and experiences" of ethnically different students (Gay 2000, 33). This runs counter to many teachers' feelings that taking a "color blind" approach to teaching enables them to treat all students equally (Irvine 2003). These teachers' concern is that in varying instruction based on ethnicity they are enacting racist or biased methods. Ironically, in treating all students the same, teachers are enacting the very prejudices that they fear the most. One mode of instruction does not serve all students in the same way, and inherently favors one group. This is most likely to be the group that is most like the teacher, as the teacher is likely to act out of his or her own cultural experiences. Since most of the teaching force is white, the students in this group are most likely to benefit.

Culturally responsive teachers validate and affirm student cultures by recognizing them as such in the classroom and paying tribute to each of them equally. These educators teach through the strengths of the student, "build[ing] bridges of meaningfulness between home and school experiences" (Gay 2000, 29). In this way, CRT is multidimensional and comprehensive, encompassing many aspects of student and community lives. This validation and value placed on student lives is powerful and liberating.

Culturally responsive teachers look to transform individual perceptions and actions through education. Transformative teaching involves helping "students to develop the knowledge, skills, and values needed to become social critics who can make reflective decisions and implement their decisions in effective personal, social, political, and economic action" (Banks 1991, 131). In this way, CRT is emancipatory and transformative for students, teachers, families, communities, and schools, re-defining what is considered important knowledge, validating students' every day experiences (Ladson-Billings 1994), and engaging in a continuous effort to combat the status quo. Student cultures and familiarities become a part of the curriculum, and the norms of the classroom are re-characterized based thereon. Specifically, modes of communication are made to benefit each child, ways of assessment and teaching are re-conceptualized, and a trusting and caring environment is emerged. This empowers students to believe that they can succeed in learning tasks, and motivates them to continue meeting challenges in and out of the classroom.

Culturally responsive teachers recognize and utilize the fact that other forces are at work in schools, and that pedagogy, though a powerful tool, cannot alone change societal perceptions (Gay 2000). For example, culturally responsive teachers recognize that academic achievement may be frowned upon in particular cultural groups because of historical and social institutions. As a result, academic knowledge and attainment may be masked by a student as a defense against alienation from his or her cultural group (Fordham 1993, 1996). Again, this is illustrative of a severe discrepancy between student and school culture. The negative views of education held by certain cultural groups are understandable; the educational system has consistently asked students of color to abandon their culture for the sake of education (Nieto 2002). It makes sense that a delineation separating the two would emerge as a defense mechanism, a tool for resistance, or a way of simply denying what the school culture deems important. This undercurrent of cultural friction implies a school culture and, more widely, a societal issue relating to the nation's view of education, and education's view of the nation.

Culturally responsive educators aim to change the face of education, helping students to realize that they can control their own learning. Cultural beliefs and practices are at the forefront of instruction, and are viewed as individual strengths rather than collective weaknesses. In this way, culturally responsive teachers aim to transform traditional educational practices and negative views of ethnic students of color (Gay 2000). As such, CRT cannot achieve transformation, validation, and empowerment in isolation; rather, it must be a part of a larger school and community change in perspective. As such, specific culturally responsive teaching practices vary widely between and among schools.

In bringing student knowledge and experience to the forefront of teaching, culturally responsive educators aim to develop students' intellectual, political, social, emotional, and cultural understanding. As such, CRT requires that teachers consider many factors within the educational setting, including curricula, learning environment, assessment techniques, patterns of communication, and relationship development. As such, culturally responsive teachers aim to educate holistically, allowing students to maintain their ethnic identities while developing a sense of shared responsibility in the learning environment (Gay 2000).

The multidimensional nature of CRT requires that it, and the teachers who enact it, are flexible and continuously evolving. If CRT were stagnant and rigid, it would inherently contradict its own purpose. Thus, perhaps the most important and distinctive characteristic of CRT is its flexible nature. Indeed, CRT will not look the same in any classroom, and the enactment of the framework may vary widely. It is through this continual evolvement that CRT can change with the diversifying population. Moreover, it is only through adaptable methods that educators can hope to meet the needs of various learners.

Culturally Responsive Mathematics Teaching (CRMT)

The framework of CRT has remained somewhat general in terms of subject-specific strategies and theoretical foundations. Certainly, each subject should not be separated from the others as this would simply lead to more topical, cookie-cutter solutions. The current framework of CRT, however, serves as a context within which culturally-based mathematics teaching can be examined. In other words, this work is not meant to be divisive, but is meant to zoom in on one specific subject area within the larger context of CRT. In adding to current ideas and revering successful models, this work becomes more prevalent and relevant to schools.

Through interviews with Gloria, observations of her classroom, collection of artifacts such as teacher-made tests and student work, and informal conversations that occurred naturally, the theory represented in figure 4-1[2] was developed in an effort to generalize her practice into a working premise for culturally responsive mathematics teaching (CRMT). This model was constructed inductively using grounded theory methodology (Glaser & Strauss 1967), and only data collected in Gloria's classroom (interviews, observations, artifacts, informal conversations) was used in the analysis that led to these conclusions. Given the limited research base (there are very few research based discussions of the construct) in the area of CRMT, grounded theory was an appropriate methodological choice. Subsequent interviews with other persons that are featured throughout this book have corroborated these findings, so all of these voices are useful in describing Gloria's unique pedagogy.

Unlike traditional, hypothesis-driven, deductive theory building, grounded theory aims to systematically generate a general theory from specific data (Glaser 1992). While the former is based on the testing of a predetermined hypothesis, the latter is developed with no presupposition and is generated as data are collected. In this case, the specific data are the interviews, observations, and artifacts mentioned above. It is important to note that in my work I treated all interactions (formal and informal), conversations, relationships, and perceptions of those involved as data. In doing so, I became the lens through which all data was sifted. Though a systematic process was followed in sorting data, choosing what was important, and emerging the following theory, subjectivity is inherent in its development. The theory, however, is fully grounded in the data.

As is implied by the choice of grounded theory as a methodology, this work is not meant to pigeonhole or simplify Gloria's teaching in any way, nor is it meant to fully describe every aspect of her instruction. Indeed, the complexities of her classroom cannot possibly be captured in such a way, and the goal of this work is not to simply give guidelines and formulas so that others can duplicate particular practices. As such, figure 4-1 is not meant to be a grand theory; rather, the working theory is meant to serve as a substantive theory that is useful to practice because of its rooting in naturally occurring experiences. My main goal in presenting this theory, then, is to identify general, key aspects of her teaching and describe how such elements work together to create success in mathematics for underserved students.

It is important to point out one major assumption that is inherent in this work relating to teacher beliefs. CRT and CRMT are seemingly impossible to enact without a teacher's true and unwavering belief in the children with whom they work. As a culturally responsive teacher, it is not enough to assume that you are not racist or do not hold skewed versions of reality based on the status quo. Rather, personal assumptions must be examined and the educator must truly believe that children can excel in mathematics without qualifiers. This type of belief in students was perhaps the most significant force that drove Gloria's work. It is not part of the figure because beliefs are tacit. Though they change with knowledge and communication, they must be in place to ensure success in the classroom.

Figure 4-1 represents the dynamic and fluid nature of the teaching practices of Gloria and is meant to highlight the complex and fluid interactions that were inherent in her mathematics teaching. There are cornerstones, represented as the "walls" of the model, and the cycle of pedagogy and discipline exist within and are influenced by these ramparts. The student is at the center of all interactions and efforts and there is continual movement around the student.

There are, in fact, four cornerstones of Gloria's instruction represented. Three are more obvious: knowledge, relationships/trust, and communication. The fourth, constant revision and reflection, is represented in numerous ways, the main of which is the arrows between the three others in the model. These cornerstones are the foundations of Gloria's work, and serve as the platform from which all of her success was possible. Strength and focus in these four areas were foundational to her mathematics instruction, serving as the bases from which lessons, interactions, attitudes, and pedagogy were constructed.

Central to the figure is the student who is constantly involved in the cycle of pedagogy and discipline. The pedagogy is always moving, in a sense, with unannounced jumps between concepts, varied teaching styles, and new material occurring at any time during her mathematics lessons. Moreover, pedagogy and discipline are so closely tied that one cannot be distinguished from the other. Her pedagogical and disciplinary actions were fluid and connected, and behavior was often controlled through pedagogy. Still driving the decisions made in this cycle are the cornerstones, which are always present, stabilizing the student in learning mathematics.

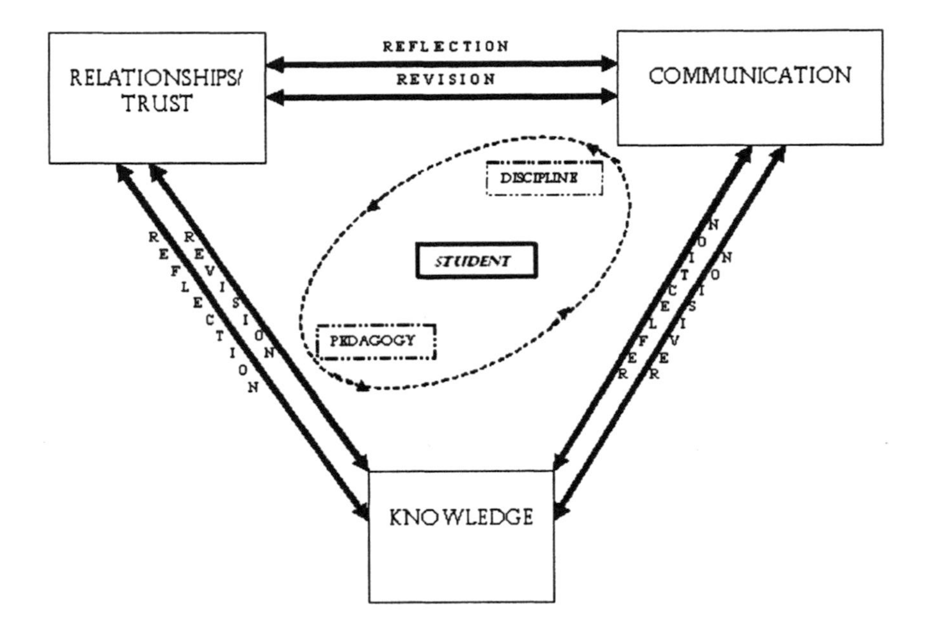

Figure 4-1. Culturally Responsive Mathematics Teaching (CRMT)

As the arrows also indicate, the cornerstones are not mutually exclusive, each informs the others, and they could not be in place without the others. Many communication techniques such as chants, choral responses, and direct, energetic instruction, for example, are directly related to the knowledge Gloria has of her students. As such, they should not be taken independently. Further, within these foundational pieces are the intricacies of pedagogy and discipline, which are constantly intertwined and enacted continuously.

Cornerstones and Interactions

Each cornerstone has some unique qualities, but is most useful and interesting when considered in the context of the other three. These cornerstones were the foundation and support that drove Gloria's instruction, helping her to establish a strict, academically focused, and caring environment. An explanation of each is given below, and it will be obvious why each cannot be considered in isolation. Several interactions and outcomes of these foundational pieces are explored in the latter half of the section.

Knowledge

Having 30 years of experience teaching at Duval, Gloria had a great deal of knowledge in many areas, and used all of this information together when designing techniques and lessons for mathematics teaching. As evidenced by her explanations of content and ability to improvise, Gloria had a great deal of content knowledge and pedagogical content knowledge. That is, she knew the mathematics that she was teaching at a deep level, allowing her to analyze student errors, deconstruct concepts, and present concepts in an accessible way. Moreover, this content knowledge was applied continuously in a teaching context, making it possible for Gloria to activate her mathematical knowledge in the course of teaching (Ball & Bass 2000) in appropriate ways that allowed for greater student understanding of mathematical concepts.

This strong knowledge base allowed Gloria to maintain a very academic focus her classroom with the ideas of accessibility and empowerment as key elements in her mathematics instruction. In order to mobilize this knowledge, however, she needed a great deal of knowledge about her students. Angie remembers Gloria's commitment to learning about her students. "She spent a lot of time understanding their language, understanding their music, so she knew what it took to captivate them and get their undivided attention. It wasn't a fly by night process; it was something she thoroughly engaged herself in."

In order to gain this information, Gloria was in constant communication with students about their current interests, family lives, goals, and dreams. Moreover, she was an active member of the community and spoke with parents, formally and informally, on a continual basis. With this knowledge as a base, she worked continuously to design techniques that tap into student interests at a cultural level. Observation of students, as implied by the quote at the top of this chapter, was the main way in which she attained this information. "She paid close attention to her kids and she knew their needs as far as intellect, emotionally, family support," recalls Angie. "It takes a special person to do that."

Communication

In order to effectively present mathematics in ways that made sense to students, Gloria relied on several modes of communication. Largely based in cultural practice, she carried familiar communication patterns into the classroom. This was especially effective because such patterns were also familiar to Gloria. In particular, she used call and response techniques integrated with rhythms. For example, at the end of each problem involving a fraction (which was often solved in unison), she would chant "question" while snapping to which students would respond, "Is it in simplest form?" This back and forth would repeat several times as the students studied the problem while chanting. After two or three repetitions, Gloria would keep the beat and say, "Answer," to which the students would respond, "No." This "answer—no" pattern would usually repeat two or three times. The students would then go on to reduce the fraction in unison using a similar technique.

Cultural communication patterns were also evident when Gloria took disciplinary action in the classroom. "It's a cultural thing I think," she would say. "White kids learn differently…The parents [of my students] don't talk to the kids the same way [that white teachers do]. The majority of [African American] parents do not say 'okay now I'm gonna give you one more turn or you're gonna be in time out.' They say, 'Johnny if you don't go sit down, somewhere, you and I are gonna have it!' Once [teachers in the school] figure out these culture things, the kids work out fine." Though her tone at times seemed harsh, students understood that it was Gloria's belief in their ability and high expectations that drove her actions.

Relationships/Trust

Gloria was a teacher in the same community for nearly 30 years. Over time and after much success, she had built strong relationships with families in the neighborhood, the school community, and the students that the school served. Her no-nonsense attitude in the classroom and work that went beyond the school day earned her the trust of the community and the students. Moreover, her history as a member of the community allowed her to share the perspective of many of her students and their families—this was obvious in her commitment to the students. "When you have built that kind of trust," recalls Lee, "and the caring that the children knew, you have the children eating out your hand. You've got them…hook, line, and sinker right there. And I think that's what she did best."

Constant Reflection and Revision

The reflection and revision process was so quick and so constant in Gloria's teaching that it was sometimes hard to identify. When I would speak with her after a lesson, she would often use the conversation to speak about the technique, what worked, and what didn't work, eventually deciding on ways to make the lesson more relevant and engaging. Many times, this type of process would occur during the lesson. Gloria had a way to feel the pulse of the group and make adjustments as the lesson progressed. If she felt that the students did not understand a concept, she would stop and probe the class for misconceptions. She would explain the concept different ways, perhaps trying a new technique or cultural connection, until the students were able to connect to the material. From year to year such revisions and reflections were also common. For example, in the 2005-2006 school year, students were generally not allowed out of their seat. By the 2007-2008 school year, students were at the board working problems on an almost daily basis. "I always change," she would say. "I get tired of the same thing over and over, so it's always time for me to change and get to something else."

This reflection process is key to mathematics teaching and learning, and "teachers must be willing to change." This is evidenced in specific teaching practices. Gloria, for example, acknowledged not only that she wanted to change, but that she was "at that point in time where things have changed and I think I need to change my technique" so that more individual accountability is required of students. Also, she constantly changed cultural pieces, such as the music she used to teach mathematics concepts, in an effort to stay current and maintain student interest. Further, students inform the reflection process, with Gloria constantly asking herself, "okay let me see what I can come up with so that these kids can understand the different [concepts]"; this type of student-centered mathematics instruction cannot be effectively enacted without an effective, culturally responsive learning community.

Cycle of Pedagogy and Discipline and Culturally Responsive Teaching Methods

Together, these cornerstones were the foundation of Gloria's work. Though evidenced in her teaching, they were also cornerstones in her life, influencing her actions, interactions, ways of being, and attitudes. Within these strongholds was the cycle of pedagogy and discipline that specifically took place in the classroom. The negotiation of pedagogy and discipline was almost indistinguishable. Gloria's lessons were so engaging and so reflective of her expectations of students and belief in their ability, discipline was almost inherent. Moreover, students were expected to move quickly from concept to concept, often responding in unison, dancing, or chanting call and response methods, that there was little time to act out. This constant action is represented by the arrows in the interior of figure 4-1. If a student did act out, Gloria handled it quickly and directly, asking the student to stop talking or, in more severe cases, leave the classroom. All of the cornerstones shaped these types of interactions, thus the cycle exists within them.

Another way to look at the model is to consider the placement of each component. The student is the focal point and is always central to the instruction. The student is most directly affected by the cycle of pedagogy and discipline and is, in a sense, caught in the movement and relevance of the classroom techniques represented in the cycle. Though less obvious, the student is also affected (consciously and subconsciously) by the cornerstones and Gloria's beliefs that shaped these foundations. As the outermost part of the model, these serve as strongholds that work together to keep the student as the central model. If a student has a difficult evening at home, for example, and comes in the next day distracted, he may slip out of the cycle of pedagogy, figuratively speaking. The cornerstones, however, will catch the student at a foundational level, as expectations of him and beliefs about him are communicated, thus keeping him from abandoning school altogether. Often, Gloria quickly re-engaged the student. Specific examples of this type of instruction involve communal learning, rhythmic teaching, improvisation, and a constant focus on review. These culturally responsive teaching methods provide familiar contexts for students and increased achievement.

Communal Learning

The learning context is central to academic success for students of color. African American students have been shown to perform at higher levels on mathematical-estimation tasks when content is taught in a highly communal, collaborative setting (Boykin 1994; Hurley, Boykin, & Allen 2005). Gloria drew on this shared value, adopting a particular style that involved a lot of call and response, rhythmic strategies, peer teaching, and chants in unison. In many ways, the feel of the classroom mimicked that of a community church. There was not an overtly religious tone, but the style was similar. Students were taught as a group, for example, and were expected to participate fully. If Gloria felt the energy in the room decline, she would demand attention, using phrases such as, "y'all aren't with me, now! Get with me now!" The students would respond with "yeah!" and "we're with you!" letting her know that they were engaged and ready to learn. This was one major way that Gloria used shared cultural practices in the course of teaching. As Buffy Bondy recalls, "The music of church and just the movement and intensity of it that was all part of what she brought in—that's not pop culture stuff, that's deeper cultural stuff that she shared with her students and it became part of how they did their work together."

Rhythmic Teaching

African American students perform better on specific tasks when taught in a context that allows for and encourages a high amount of movement and music (Allen & Boykin 1991). It has been suggested that this type of opportunity to engage in various modes of learning is helpful to African American children because it may align with historical cultural practices. This interconnection of culture and cognition is believed to lead to academic success and empowerment (Allen & Boykin 1992). Gloria's use of music, chants, and specific memory techniques allowed her students to make this connection between culture and mathematics. Gloria explained:

> I have always done a lot of, you know, rhythmic - I guess - teaching. Rhythmic all the time! When I taught the younger kids, I was doing a lot with vowels and the vowel songs, and I used to hit the floor all the time [motions like hitting the floor with a yard stick]. The kids really respond to that so I do it with the math—its working… Now they get into it. The kids hit the desk like when I say "how many quarts in 5 gallons?" they say "20" and hit the desk in rhythm, then say "20" and do it again. Yeah, they get into it with that type of problem…it builds their self-esteem because they see that connection and yes you can learn, yes it is easy.

Gloria also made use of student interest in dance. This not only helped them to make a physical connection to the mathematics, but also allowed for more culturally based music and rhythm to be incorporated into the lesson. Gloria also used these connections as informal assessment tools. She shared:

> The kids love to dance and if you can put math into action they can basically retain the information just by doin,' develop by movin,' and you know kids just love to dance and they can recall the information... a good example is on a test when you might talk about a right angle you taught them the dance to learn right angle [puts arms in an "L" to illustrate a right angle] they be sitting there trying to make the right angle while they take the test. The dance really help the kids. It helps them to kinda exert their energy too at some point in time during the class time we have to get up and just say hey we need to . . . let go. And I'm involved in that too...We started something else new where the kids have to write on the white boards, and if I say "okay give me a mixed number" they have to...write a definition for the mixed number on the board and we have the music. So once they finish they get up and just do a little rock [back and forth in rhythm in their chair] . . . and all the kids are involved and by THAT you can look around at a classroom and see who understand what a right angle is . . . because you have total class participation.

These types of activities are illustrative of Gloria's "show me" technique. When students would give an answer or define a concept, she would require that they show that they understand what they are doing from a mathematical perspective. This helped students to make connections mentally about a concept, what it might look like when illustrated, and its concise definition. Certainly, this focus on making sense of information while requiring that students use proper mathematical language and techniques was one way that high expectations were set in Gloria's classroom. "The rappin and the singin' and the tappin'...brought such excitement to [the students]," recalls Leon Marshall. "Everyone felt they could do that...and she would tell them when you're walking along, you're gonna hear that beat and just say it cause it's there, like a song from the radio."

In later years, Gloria began calling on students to create their own beats, songs, and dances. Leon recalls that she would indicate that she wanted a new beat for a particular set of vocabulary words. When someone made a suggestion, "she would look at it and say, okay let's try it out...everybody try what he's doing." This gave children ownership of the material and helped to integrate their artistic talents into the mathematics curriculum.

Storytelling

Gloria always used stories in her mathematics teaching to ensure that children could relate to the material. This was key to her success, as many children had a difficult time seeing how the mathematics related to them personally. These stories were sometimes meant to demonstrate meaning behind the material. For example, when teaching children how to find the least common multiple of two numbers, she would discuss relationships and family. "Think about the numbers 5 and 10," she would say. "If 10 is the big brother and 5 is the little brother, who are you going to see when you are walking down the street?" The children understood that 10, or big brother, was the larger, more noticeable number. "Will 10 ever be like 5 again? No, because the big brother has to take care of the little brother," she said, relating to the fact that in Duval's community, older siblings often care for the younger ones. "But 5 will be like 10, right?" she would say, making the connection that the two are related so the little brother looks up to the older brother. "Now how do you know they are related at all?" she would push. "Because when you count by 5's you say 10" the children would respond, understanding that even though the numbers might not have the same name, they are in the same number family.

A similar strategy was used when Gloria would introduce or review the concept of borrowing (trading in subtracting). "If you are cooking with your mama," Gloria would say, "and she runs out of eggs or sugar, what you gonna do?" Students understood that often their parent would send them to the neighbor's house to borrow more of whatever it is that they need. "Then you have a little bit more, and they have a little bit less," eventually connecting this to the idea of "borrow 10 more," a concept of particular importance in subtraction.

Other times, stories were told as a way to help students remember facts or ways of doing problems. When doing problems that required students to use the order of operations, for example, she would connect "cleaning up the problem" to "cleaning up the house." First, you clean inside the house, and then you move outside and clean there. Similarly in mathematics, you first look inside the parentheses and take care of the mathematics inside, then you move outside and work there. These types of ideas seemed to be daily occurrences in Gloria's classroom, with new stories emerging continuously. Often, she would ask students for ways that they remembered concepts, helping them to come up with their own strategies and stories.

Focus on Review

Review was woven throughout Gloria's daily teaching methods. She rarely discussed review without mentioning music, dance, chants, improvisation, or a combination of these techniques. This focus, however, was truly unique to her classroom. While many teachers may do a quick review at the beginning of class before moving on to a new concept, Gloria would constantly go back to old material, and she would do so on a moment to moment basis (which ties in closely to improvisation). "Every day was an opportunity to go all the way back to the beginning and review everything you know, and that is not how I tend to see people teach mathematics," says Buffy. "They might do a minute on what happened yesterday, or if its Monday maybe something about what we did last week, but hers was this ongoing review."

Since definitions are learned communally and are communicated orally to the teacher in a call-response fashion as a class, Gloria would ask for them at any point in the class. She learned that this type of repetition worked for her students by simply paying attention to student interests. By tying the definitions, and therefore review, in with music or communal learning, students were able to retain information over time. Gloria once said:

> At the beginning of the year I found my kids just comin' in and singin,' singin,' even around my house with my little grandbaby is 3 years old and I noticed that she can sing a song over and over, memorizing all these words. So we can't say kids can't learn, you know, she's 3 years old, so I said well let me come up with this idea. These kids understand THROUGH MUSIC, and through repetition, the kids learn it… I thought about Star Spangled Banner - we didn't learn it in one day, we had to keep goin' over and over. So the music plays an important part with our kids, simply because our kids look at BET, and everything is on BET for them, and they strive just to look at BET so I am at that point now where I will do my lesson off of BET…you find some of those kids look at BET for hours…because they enjoy the music. I said… okay, so I can do the same thing for math . . . because it has to be repetition. We go over and over so now it just falls into place, and they understand. Now I call it rapping with math, and it is rapping with math, but do you understand what you are rapping about? That's very important that they understand anybody can rap, but when I say perimeter do you understand what perimeter is? You know, anybody can just rap a song. Then I tell them too, even the song that you listen to every day, do you understand what those songs mean? It's very important that you understand the meaning of song. [The math rap is] a lot of um definitions incorporated into just the singin.'

Improvisation

With all that was going on in a given day in Gloria's classroom, it might be difficult to identify a specific teaching plan for the day or any sort of prescriptive lesson. "This is not someone who developed lesson plans," recalls Buffy. "So people who… would try to figure out what exactly is she doing could be a little bit mystified. It was like she sort of felt the pulse of the group and would just blast off in any number of directions and then, almost like the old pinball machines she'd just zip, zip, zip from one point to the next." This was evident in every aspect of her teaching. With review, for example, she may begin the class by asking for the definition of a mixed number. If the class responded confidently, she might immediately begin a chant asking students to convert quarts to gallons in rhythm. If the response was solid, she might draw a figure on the board and ask them to find the area or perimeter. If she sensed a lack of confidence or hesitation in any individual, she would stop and drill. At any point, she might jump into new material. If the kids were not with her, she might change directions. As Buffy Bondy remembers, "her lesson plans just evolved in the process of teaching and watching how kids were responding to her."

Warm Demander Pedagogy

Particular types of pedagogy have been shown to be effective in bridging this school-home gap for African American students. Teachers who enact such pedagogies maintain a firm, directive, and demanding demeanor while communicating a deep sense of care for students. African American students have been shown to have positive responses to this type of "warm demander pedagogy" (Ware 2006). Warm demanders are explicit and clear about their expectations, exhibiting an ethic of caring, positive beliefs about students and the community, and specific instructional practices, all of which create a "positive psychological environment" (Ross, Bondy, Gallingane, & Hambacher 2008, 143). These teachers show respect for students and culture, and are "characterized by respectful interactions, a calm tome, minimal student resistance, and a clear academic focus" (142).

Students view warm demanders as authority figures and disciplinarians, caregivers, and pedagogues. In research involving warm demanders, various examples of direct instruction and inquiry learning have been identified, as have specific examples of culturally relevant pedagogy and classroom management. Overall, this implies that these educators' classrooms and teaching practices are multifaceted and complex, the teacher taking on many roles in an effort to push students academically.

This sort of "tough love" has been shown to have success with African American populations for decades. Even before integration in schools was mandated, successful African American teachers were "consistently remembered for their high expectations for student success, for their dedication, and for their demanding teaching style" (Siddle Walker 2000, 264-265). This type of work is especially important for young, white women, the best-represented population in the teaching force. White teachers have often "been socialized to speak softly and to be non-direct and non-assertive" (Bondy, Ross, Gallingane, & Hambacher 2007, 146), making them less inclined to use a seemingly harsh tone with children. In a largely African American setting, these attributes may communicate a lack of authority and weakness in discipline.

"[Gloria] is the perfect example of a warm demander," says Buffy. "She communicates how committed she is to them and how much she cares about them from the beginning so then when she has to get tough, which she will not hesitate to do, they respond to that because they know that this is someone who has their very best interests at heart and really wants to see them excel." The following examples from Gloria's classroom illustrate this idea well. In the first example, Jeffrey[3] is a student in Gloria's classroom. She had struggled on and off. This exchange took place in March, towards the end of the school year.

G: Jeffrey, where is your math journal?
J: [inaudible mumbling]
G: Where—is—your—math—journal?
J: it's....
G: [interrupting] Boy, look at my face when you talk to me and push that chair

in.

J: It's at home

G: It don't do much good when it's sittin' at home. I don't know why I even bother if you can't do the work. Get up out your seat and go call somebody.

J: I got nobody to call.

G: Get yourself UP out of your seat and call someone—they can either bring your journal or bring themself and take you home. Can't come to my class unprepared.

The second example is illustrative of the type of force with which students were met if they were not on-task. Many variations of this type of interaction were a regular occurrence. In parts of this example, the students are responding in unison as a whole class (denoted by S) or by group (denoted by SG) during the course of a normal lesson. Students are seated in groups of 4 or 5 and there are 45 students in the classroom. They are working on changing mixed numbers to improper fractions using music and chants. One student, Brittany, is eventually singled out.

G: [starting music and snapping to the beat, points to the first in a line of mixed numbers on the board, 5 ½] Go!

S: [in rhythm] 5 times 2 equals 10 plus 1 equals 11 over 2. 11 over 2 [Gloria points to the next problem, 3 ¾]. 3 times 4 equals 12 plus 3 equals 15 over 4. 15 over 4.

G: [points to one group with one hand and the next problem, 2 $^{3}/_{5}$, with the other]. This group, go!

SG1: 2 times 5 equals 10 plus 3 equals 13 over 5. 13 over 5.

G: [pointing to next problem, 4 $^{2}/_{3}$, and next group] Go!

SG2: 4 times 3 equals 12 plus 2 equals 14 over 3. 14 over 3.

G: Stop, stop, stop! [stops the music] Brittany, you just moving your mouth, baby, ain't nothing coming out.

B: Yes I am!

G: [turns the music back on, resumes snapping to the beat, and points to the 4 $^{2}/_{3}$ problem]. Okay. Ready, go!

B: 4 plus, uh…

G: Go ahead! [continuing to snap]

B: 4 times 3 equals 8 plus

G: 4 times 3 equals 8? You don't even know your multiplication facts.

B: No…it's 12…

G: Look back here child. See this big piece of paper? [pointing to a large piece of poster paper]. During lunch your assignment is to write multiplication facts until you can't write no more. [walks back over, turns on the music, and resumes the whole-class lesson]

To outsiders, Gloria's tone of these exchanges may seem harsh or demeaning to the child. I must admit that I was taken aback the first time that I heard her use this tone in her classroom. Buffy recalls similarly that "people who observ[ed] her might think 'oh my, she was a little harsh with that child' or 'that's kind of a tough tone she's using, these are just 5th graders'…I must admit that when I first saw her I thought she was a little...scary." The children, however, consistently responded to her no-nonsense approach and, as was evidenced by the stories at her wake, believed that she was tough because she cared about them, wanted them to excel, and believed that they could do so. This is illustrative of the focus that Gloria kept on the student—outside norms and status quo practices did not guide her; rather, the particular learners in her classroom drove her instructional and disciplinary practices.

These types of interactions showed dynamic interactions between the cornerstones. Each took place in the course of a lesson, was dealt with quickly, and did not interfere with the academic focus of the classroom. Through these interactions, she communicated high expectations and care through culturally-based verbal and nonverbal cues. She did this using particular, familiar language and tones of voice which were based on the knowledge she had of her students. Moreover, Gloria's commitment to the students was evident through her communication style and attention to their interests and ways of knowing. This helped her to build strong relationships with the students, who trusted her as a caregiver and teacher. Student responses and continued successes evidence the effective nature of Gloria's demanding nature.

This work provides evidentiary support of Boykin's (1995) psychosocial integrity model. As opposed to the deficit model described by Delpit (1995), this model is based on the idea that children's realities, ways of knowing, and ways of being are complex and fluid (Boykin & Allen 2004). If student realities are disregarded in academic contexts (for example, when home languages are stifled and suppressed in school), educational achievement and outcomes of learning are compromised and students are likely to act out in resistance to the mainstream (Allen & Boykin 1992). If students are given the opportunity to act within familiar cultural contexts, however, meaningful learning is more likely to occur and accessibility to higher-order thinking skills such as problem solving is increased (Boykin & Allen 2004).

Outside Support

Gloria's involvement in the community and unwavering support of students as points of pride in her life were real factors in her success. Each year, the fifth graders at Duval look forward to many opportunities for involvement and several special events. The most talked about of these was the opportunity to be a part of the Math Team. Gloria was in charge of the organization, and any child that wished to be a part of it could try out; the total number of students varied, but was generally around 30. A try out consisted of an oral testing, using the same methods taught in class, of concepts learned during school. The result was a team of exemplary students who would showcase her methods (e.g. chants, dances, etc.). It is also important to note that these were not necessarily the students who had always been at the top of the class. Indeed, there were ESE kids on the team, though you would never know who they were. The Math Team began performing in churches and other community settings (they were usually invited), and after the word got out, they were invited to perform all over the country. In 2007 and 2008, the team was invited to Atlanta, Georgia to perform. In 2007, Gloria and her team traveled with a team of researchers from the University of Florida to present at the annual meeting of the Florida Council of Teachers of Mathematics. Invitations were also accepted from groups in Tampa, Miami, and Gainesville.

Audience members were always amazed by the performances, and often asked if we had any materials based on her teaching that they could buy. "It was a road act," says Leon. "People couldn't believe it." Angie remembers that "the Math Team became the spokesperson for what she did. She wanted to get [her mathematics teaching methods] out there in a way that people could see [them]. She felt like she could go to workshops and talk about things and do a power-point, but she thought that nothing was more revealing and compelling than bringing the kids and letting them showcase it for themselves. And she was right—people would just be amazed at how these kids could do what they could do. This became her showcase or her mouthpiece for the things she was doing in the classroom."

The Evening of Elegance, which is also only for 5th graders, is a formal event that occurs at the end of the school year. In Gloria's words, the evening consists of "a program where the kids get to wear their formal attire, and we introduce each student to their parents and family, introduce their parents if they are there, so it's something like a formal banquet." This, in addition to a graduation ceremony that also occurred at the end of the school year, was almost exclusively organized by Gloria. The events served as rites of passage for the students in Duval's community. The kids looked forward to it all year, some even buying their dress for the Evening of Elegance as early as January. This desire to be a part of these events not only motivated the children to work hard, but also gave them a reward for their work and persistence in school. In the course of the ceremonies, the pride on each student's face was evident, as was the feeling that they were somehow empowered to go out on their own carrying with them what they had learned at Duval. Gloria also wrote and directed the 5th grade play each year, parts of which were often performed at the Evening of Elegance.

These types of events took time and effort to plan, as they often included a ceremony, a large dinner, decorations, and speakers. Leaders from the community and the university were invited as special guests. Gloria often took control of the planning, mainly to make sure that each child was able to participate. With a formal event, the cost of clothing is often an issue that might prevent this. "I do have gowns available for the ones who can't afford them; we get donations from different organizations, and I go out and find gowns for the ones who are having a difficult time." Graduation posed a similar problem, so "she made all of the graduation dresses for the girls," recalls Lee McNealy. Leon Marshall recalls that she also "made all the corsages, she went out and bought all of the shirts and pants [for the boys]—every child looked the same. No child has ever been denied graduation clothes because they didn't have. We've got containers of clothes that she purchased with her own money or made just so the kids could be successful."

Gloria's willingness to reach out to families and community members as participants in her teaching was truly unique. Years ago, Gloria noticed that her students were not getting the help that they needed at home. As the confident and outspoken woman that she was, Gloria "asked [parents], what is going on at home? Why won't you help your kid with his math homework?" Essentially, she was concerned that a lack of support at home would have a negative effect on all that she was trying to accomplish at school. Rather than blame parents[4], however, she asked them what type of support they needed from her to help their student at home. "I talked to one woman when she was dropping off her kid, and she said, 'Ms. Merriex, you know I don't know that math!' So I told her what if I teach you?" As a result of this and several other conversations, Gloria began to offer night classes to parents who wanted to learn the mathematics. Some parents wanted to learn the subject for themselves and their own lives; others wanted to be able to help their children at home and realized that knowing Gloria's methods would make their child's learning more seamless. Initially, however, parent responses to this offer were met with cynicism and a bit of reluctance, but over time attendance grew. As attendance grew, student enthusiasm grew, as did the quality of work produced at home.

Gloria taught the adult classes using the same techniques that she had developed for her younger students, but without the controlling overtone that is developmentally necessary for the children. Gloria believed that this type of interaction helped her to build trusting relationships with parents. In turn, these relationships directly impacted student achievement and gave Gloria insight into how the children live and interact with their families. "[Parents] let their hair down when they are in my class," she would say, "and we have a good time. I have a good time explaining the problems to them, no put downs whatsoever. You know, we just come in here and talk and we talk about how to solve a problem and with that I have a good relationship with my parents now. It really helps the kids." In recent years, Gloria was so busy with other things, such as conferences in various cities that she would travel to with the Math Team, that she could not offer the evening courses. She had planned to pick it back up this fall.

Angie put it nicely. "In theory we talk about educating the whole child, and she totally understood that. A child can't learn if he's hungry. A child can't learn if everybody else has on clean and decent clothes and he or she doesn't—they're going to feel the impact of that. She paid attention to everything and that's what it means to educate the whole child."

Overall, these facets of Gloria's instruction and life drove her decisions in the classroom. The key components of communication, knowledge, relationships, and reflection served as supports and bases for all of her practices and commitments. The success of Gloria's instruction was most obviously evidenced through high student scores on standardized tests, students' future academic endeavors, and academic achievement of students in general. The impact of her teaching on intangible and unquantifiable aspects of students' lives, however, is long-lasting and transformative. Such results evidence and evidence the dynamic nature of the aforementioned theory of CRMT, and serve as examples of how all parts of Gloria's instruction work together to achieve phenomenal results. As students became a part of her classroom and life, their self-perceptions were changed, and their understanding of societal practices and norms became more comprehensive. Student shifts in worldview was one of the most dramatic outcomes of Gloria's work, and relates to all four cornerstones.

In particular, Gloria's work allowed students to grow as mathematicians while developing racially and culturally. In other words, students were not forced to conform to a set of norms that did not align with their home culture; rather, their culture was used as an access point to mathematics and as a framework for learning. Shared cultural values and practices were central to the classroom, and success was defined in these terms. As such, students viewed mathematics learning in a new way that was meaningful, thus coming to see themselves as able and excellent mathematicians. Moreover, mathematical knowledge came to be seen in a more positive light communally. To parents, students, and community members, mathematical knowledge was no longer stigmatized as something unattainable or for which cultural values must be sacrificed; rather, it was seen as a vehicle through which racial identity development was enhanced became a shared value among school and community members.

Part III: The Legacy of Gloria Jean Merriex

We must not legitimate the inequity that exists in the nation's schools, but attempt to delegitimate it by placing it under scrutiny. In the classrooms, working in opposition to the system is the most likely road to success for students who have been discounted and disregarded by the system.

-Gloria Ladson-Billings,
The Dreamkeepers, 1994, p.130

Chapter 5

For students, I'm not trying to reward you for something they need *to do... your reward is me standing before you trying to help you. Whatever you get out of my teaching is* ***the biggest reward*** *you can ever get. [Some teachers are] so busy now [at the end of the school year] and are getting away from staying on the kids. The kids check out. [Teachers] are so busy giving parties and giving them lollipops or points and not teaching. I'm not gonna do that. That's just me.*

-Gloria Jean Merriex

The following pages solely consist of direct transcriptions of student stories about Gloria Merriex. These narratives were gathered in December of 2008 from 5 of Gloria's past students and 3 of those students' parents. Each of the students is now (in 2009) in middle school. Courtney[1] is a 6th grader at a middle school in Alachua County. Courtney's parents, Lisa and Tim, also participated in the interview. Jennifer is a 7th grade student at a middle school in Alachua County, and she interviewed alongside her sisters who were participated with Jennifer in reciting the math rap that is transcribed below. Diana is Melissa's mother.

These students represent the impact that CRMT can have on individual lives, and are a testament to Gloria's work. Further, student and parent voices are often absent from this type of work, or are altered to fit a theory or idea. I feel that these stories do corroborate my research findings, but leave this determination up to the reader. As such, these stories are left untouched and free from interpretation.

Courtney

She taught more like by moving and singing and stuff rather than just sitting at a desk and working out of a book. Until 5th grade I never knew my times tables, but she made me learn them by singing them and it was easy. Whenever I forgot to do my homework, or anyone forgot to do their homework, they would have to line up in the hallway and talk to Dr. [Lee] McNealy, the principal, and [Ms. Merriex] would kind of yell. One time, it was kind of scary, we were getting picked up from school and my mom didn't really park the car all the way, so Ms. Merriex came in there and she was like 'This car isn't parked right—where are the keys." We said, 'Our mom has them and she's inside.' She said, 'Well tell her this car isn't parked right!' That was the day before she died. Sometimes she would just make something up on the spot because she just thought of something that came into her mind. Also sometimes the kids would do something or come up with something and she would say, 'Oh, that's good, everyone do that.' This one day, I forgot which one it was, but these kids were kind of shaking their booty doing it, and she told everybody to do it that way. She made it feel easier to learn. I felt confident about math in her classroom and all the stuff she taught us she made us write down in that journal. Sometimes when I'd be getting picked up I would just read over that journal. The thing is everything in the journal you could also sing it which helped. Even if you were really bad at math, [she showed me that] you could always get up to the level and even do it really fast sometimes. If I didn't have her last year I would be really stuck because I didn't even know how to do long division or anything. I'd be really stuck if I hadn't had her last year.

She didn't just say, 'Oh, you didn't do your homework, I don't really care, you'll just get an F.' She actually told us that 'You really need to do your homework' and made sure we did. Actually before I went to Duval, I just went to Jefferson[2], and everyone there thought it was just so cool to be stupid and dumb, but here [at Duval] people would brag all the time about, like, if they knew their times tables better than someone else or they could say them faster. At Jefferson that would be so nerdy, because at Jefferson it was so nerdy to be smart. I didn't really see everything, but most of the things that I see people bragging about were taught by Ms. Merriex.

Ms. Merriex actually focused on the kids who weren't doing well on purpose. Like she'd tell people to come up and do a problem on the board or something and she'd pick on students because she knew they needed to work on that certain thing, she'd pick on them on purpose to make them learn that. Sometimes she'd pick on people to sing the times tables by themselves, and that's actually how I learned it. At first, I would kind of mouth it and sing along with the class, but once I saw she was picking around at people, I just quickly memorized them.

When I at Jefferson in 3rd grade, which was like my worst year, I had a really bad teacher, and that's why I switched to here for 4th grade. Well, [this teacher] actually started to tell us about the math rap and Ms. Merriex and everything, but I didn't really understand it and she didn't tell us the school or anything. She just wrote the math rap on the board one day and we sang it once, and then we just stopped and she left it up there. And we actually watched the video of Ms. Merriex's students singing it, but she didn't explain really what it was so it didn't make sense.

We would perform it on the math team. Not everybody could be on the math team. You had to do your homework and stuff, but even if you forgot your homework for one time, she actually gave you like a week to turn it in, she gave you another chance. Sometimes I just forgot, and I didn't get kicked off the math team just for that reason. It was fun when you got to travel—it was kind of like missing school, but you're still learning. You didn't really have to try out—she would at first just put everybody up in front of the classroom and we'd start singing. Then she'd just like walk around and if she saw someone was playing around or not paying attention, she'd tell them to sit down. She just kept doing that until there were just a few left. But there wasn't just like one chance to be on it. Sometimes if she saw someone improve, she would tell them they could be on the Math Team. It was like, if there was a trip, she gave permission slips not to everybody, but sometimes I saw her give one to people who hadn't been on the Math Team before. After we did the math rap and everything, she'd ask what's this, this, this, and this, and she'd actually kind of like start teaching, but it was kind of like a performance. She would do different problems, but in the same way she taught. Always at the end we always sang this certain song that wasn't really math, but it kind of turned into a math thing. It was like, how Duval is good, and how it was good to be smart, and then it turned into saying your times tables.

Some people thought the whole thing was memorized, like she had taught us before that this was the answer, and this is what you're going to say. But, it was the way that she said it, it was how we were saying it back that was how we knew it, and the technique of how to get the answer too. She would give us new problems [during performances] but she made it so you could get the answer really fast using her technique, so we were solving problems on the spot. Something like if this is the improper fraction, what is the mixed number on the spot. Sometimes when I'm doing something like multiplying or if I get really stuck, I'll still use those strategies in my mind.

For graduation, we had this song that we were singing and we were just going to sing it normally like we always sing it, but after [Gloria] died they made it dedicated to her because it was a song about friendship. It seemed different without her. When we were practicing [the song] with her it seemed like the energy was really high, but when we actually did it without her it actually went down a little. She actually really increased the energy everywhere, like if people would start falling asleep in class, she would really get into it and they'd be wide awake. If everyone started getting tired, she'd make them stand up for the rest of the class. She wanted us to be loud. If we were too quiet, she would say 'I know some teachers want you to be quiet, but that's backwards.' She knew if you weren't participating and she'd make you stand up and do it by yourself, but I liked that.

Lisa

I was really impressed with what she did with the kids. So many kids just have so much trouble with math, or they get this idea that they can't do math. But she had this idea that everybody could do it and everybody could be good at it. She was really committed to that. She had this unusual method of teaching that was so effective. The body part of it [was unique]. The way if you had a ray [on a graph], you had a fist on one side, and on the other a ray kind of goes off like this. You kind of dance it out in your body so you really get a sense of infinity that way and a point here, and the angles, and the turning around. That getting it in the body was great, rather than just have it all be up here [points to head]. It's kinesthetic.

Another thing that really struck me about Ms. Merriex is that she had so much passion for what she was doing and obviously had for such a long period of time. I really admire a teacher who is always trying to struggle to find a new way to get information across rather than just sticking with something that somebody else figured out or they tried 15 years ago and are just trying out with a new batch. She was always trying to find some new way to get through to her students.

She has gone through a lot—I would come in sometimes and she would have to hold the forms right up to her face. I don't think she could really see at all, and you know, she was struggling through that and yet telling the students that they could achieve anything. But still she was so caring of all the kids, I mean she just cared so much about all of the kids. When they had her memorial service, the number of kids that just got up there—just the caring she had—I mean she made all of the graduation dresses.

She made math more cool even though she was tough. Like the way she kind of yelled wasn't a mean yelling, it was more of a caring yelling like she was upset that they didn't do something. She knew they could do better and she cared about them and she didn't want them to fail. She wanted them to do well. She was very forceful about it. She didn't say that some people were stupid and she knew that everybody could do it and everybody could do math. That's like the kind of thing, there's this idea that people get math anxiety and think 'Oh, I can't do math,' and she just said, 'Nope, none of that!'

She could really nail ya! But she was also very relaxed too with things. She just didn't carry out things in a really tense way. When you think of all the outfits she was sewing and all of the kids whose lives she was involved in and all that. She didn't seem like she was stressed out or anything, she was just very relaxed about it like she knew what she was doing. She was just a very strong character, a very strong presence.

Tim

The ability to take a risk as a teacher is a great thing, and she extended the same kind of view to other teachers that she did to the students that anybody could do it. There was one of those films that had been made when she took [the Math Team] to Orlando, and she's doing her thing with the students then turns to the teachers and says, 'It's your turn!' The teachers were silent and had no idea what to do. That was fine for them to watch the Math Team do it, but not us. So getting the teachers to jump out of their comfort zone. So many times when you see a boring classroom it is because the teacher is so in their comfort zone—how interesting can it possibly be? It's like, get up, move around, try something new, get excited, you know the same thing that the students are afraid to show. You know the students are afraid to show you that they are excited and that they care. They say, 'I'll stay back in that safe place' or whatever, and if the teacher can't cross that boundary, how do we expect the students to? Why should they lay it on the line? Otherwise the teacher stays in that judging place and students are in that judged place—is it any wonder why you want to get home?

So many students said they looked forward to going to her class. I think it's tricky though because somebody else could be doing something that looks the same and it not come across the same way. I say that because the way that she took the Math Team all over the place to teach and demonstrate what she was doing, and there was a very high standard to go on the trip. You had to have your uniform, you had to have not missed school, and turned in your homework, and it was a privilege to go on those trips even though it was also the way to teach the program. I think it made the students realize that teaching was really important and that they could be part of that, and they were included in that. And the caring somehow came along with that package deal, like, we care not just about you but that other kids elsewhere are going to be able to get to do this because you all know this is not the normal way of learning in other schools, you know. You can be a part of conveying that. I can easily see somebody else doing the same thing and it not coming across that way, so there is something intangible in there.

The other thing is that there is a lot of talk about integrating the arts with learning, and I think Ms. Merriex probably, because she was at Duval for so long as it was trying to become an arts-infused school, had something to do with the fact that the arts actually did get integrated here, because I think it's really hard. I think it's a really good idea, but the arts often do get left as a separate thing, and when it's not working, you say well, let's just get back to the teaching. In the school board and everything the arts are still funded separately. [the school board is] like, oh, well we're taking your art people out. And Duval's like 'No, wait, they're part of the core curriculum!' And they're like 'well, not actually as far as we're concerned.' And I think the solution to that quandary is people like Ms. Merriex who made it one thing. You know, you couldn't yank it out. There she had it in one classroom, and I hope and think that was inspirational to other teachers to try it, because there was an example of how to really do it, because no teacher was actually trained in this. I don't mean just for math programs, how to actually have an arts-infused core curriculum classroom is not taught in the education schools, its not taught in college, it's not anything anyone really experienced as a student, so to have a real live example.

[The Duval faculty and administration] really were committed to making [Duval] a school for this neighborhood. They thought 'you are the kids that are here. You are the kids that matter. we are going to do what works best for you and best for us.' And that was a huge part of [Gloria's] philosophy. 'this is your music, this is your movement, this is how we're gonna get there. You matter.' You know, Ms. Merriex was from here and that played into that mentality.

Jennifer

She was a special teacher. Her math made her special. The way she taught math and rapped her math. She used the rap to help us understand the math. She danced it out so we could understand what we were doing. It was fun but serious. I remember some of the math rap. You remember?

[In unison and in time with her two sisters]: Perimeter, distance around the figure. Area, length times width. Uh-huh, uh-huh! Volume, length times width times height. Hexagon, a-six sides, Pentagon five sides, quadrilateral four sides, equilateral all sides equal. Get ready, get ready for benchmark percent. A percent that's commonly used. A percent that's commonly used. Mean, the sum of numbers divided by the group of numbers. Mode the number that occurs most often. Median, middle number. Octagon, eight sides. Ninety degrees, right angle; less than ninety, acute; more than ninety, obtuse; a hundred and eighty, line. Three sixty, circle; translation, slide; rotation, turn; reflection, flip. Cardinal, in order...[laughs]...that's all for now!

I still use some of [the math rap] in class now. Like mixed number, improper fractions. I still have my journal. Some days during after school she had kids in her math class so she could tutor them and that helped us. And the strategies, like...triangle, I remember those real good. A lot of it was in the math rap.

We had to know the math really good to be on the Math Team. She chose who could be on Math Team, it depended on grades on tests and stuff, but anyone could be on it. We went to Orlando, Tampa, Atlanta, where else? Tallahassee? Yeah. We rehearsed for them, but didn't memorize the thing. She did what she did in class. She was hard but in a good way. She expected a lot from us. She cared. Her math was fun, I remember that, but she was serious about it. I remember the big sheet. If you didn't know something, you had to fill out a big [poster-sized] piece of paper writing it over and over.

Diana

She was a great inspiration and a great math teacher. The way she taught the children math, they understood it, they learned it, and they're actually using that math today. She had a personality plus, she was almost like a drill sergeant, but it was good for the children and they actually learned. The math rap, reading rap, just the different ideas and type of music she used. It's hip-hop, soul, you know it's a combination and its music that they children listened to on a daily basis. She could actually take that music and make it into something positive for the children when they were learning.

All three of my girls were on her Math Team and they really enjoyed the math. It was amazing, it really was. You'd drive up and see her standing in the back or the front of the school and she's always pleasant. The children really look forward to attending her class every day. Even on weekends if she had Saturday school, they would look forward to being here with her. It was just a learning experience to have her as a teacher, a mentor...even for the parents because we would learn math from her. You know, for the math that these kids were learning, some of us needed a brush up course and she would actually help us understand the math that she was teaching. She was planning a [formal] parent class, but we never got the chance to do it.

I actually went to Atlanta [with the Math Team] as a chaperone, what a month or two before her passing? They performed for a group at Georgia State University. These kids rapped math for an hour. That was my first time really seeing the children perform, and it actually brought tears to the eyes of some of the parents. It was just absolutely wonderful. The teachers that were at the conference, they really enjoyed that. I mean, the way she worked it was just a miracle, you know? We still have a recording of that performance in Atlanta, which was the last performance, and every now and then we pull it out and watch it. We'll cherish it forever. And it was always something different. Sometimes you would hear them perform one thing, then the next time it was something else. Things like percent—the stories she would tell to get to the answer. The kids were just into it. Class was like that too—it was fun, they looked forward to class. It was always something new.

She was an awesome individual. She just had her own style, and it worked for a lot of children. To see where mine started and where they ended up, it had a lot to do with her—a whole lot. My girls were always quiet, but Ms. Merriex would always get them to open up. She just had this thing about her. She can't be replaced. I hope somebody can continue on with her math, I really do, because the kids knowing the math—they need it every day to survive.

[My girls] still have their journal and it's like a bible. She gave them each a journal, and believe it or not, they can always look back and find something they can use in middle school. It's so funny because sometimes they come home and they're like, 'Mama, we already know this, Ms. Merriex taught us this'! I'm like, 'alright'! Thank you Ms. Merriex. Sometimes I tell them to just think about some of the things she taught you and use it, and once they work it out in their head, they think, 'Okay, this is right' and it all blends together. I tell ya, we miss her, we really do.

It was so funny because sometimes you'd run into [Gloria] out in public and she'd come up with a math question. And [the students] were supposed to have an answer just like that! She expected that [laughs]. It was funny, you know, but they'd come up with the answer, they'd come out with it. So they were really learning. So even though she was hard on them, it paid the price. It did. Like I say, they can use that math today. Whatever she taught them they can use today. That was an everyday thing. As soon as you stepped in her class, you knew what you needed to do, and she had order from day one. She let them know who she was, she did not tolerate and she was down to business, and the children knew it. Respect was there, and hey, they knew it. She kept the kids in line, she really did. No, no, no nonsense. You better not be bad when she's out of there—she comes back and finds out her class was not all business? They were gonna pay for it, you better believe it. On one of the trips on the bus, she said, 'I want everyone to get out their math journal and start reading,' and a couple of students didn't have their book. 'If you don't have one you better pretend you do'! Only once, she did not have to repeat herself.

Every parent wanted their child to be in Ms. Merriex's class, but it couldn't happen, it just couldn't. By word of mouth, each year everyone knew about the Math Team and what Ms. Merriex was doing in her class, and how tough Ms. Merriex was. But, everyone wanted to be in Ms. Merriex's class, so they knew what she was all about. I trust her with my children period. With their education, their well-being. When I dropped them at school I knew what they were coming to do. And they could show me what they were doing in school. I knew what type of person she was...just from her standing out in front of the school waving, you know, you knew that was her, and the kids did too. They listened to her, no matter what it was, and if you had a problem, you could go to her and she would help them. With anything. If somebody did not have pants, she made sure they had some. She went beyond the call of duty as a teacher. If a child needed to go home, she would take them home if they needed a ride. She was just all-around.

Everyone was so excited for graduation, you know this was her thing and she always planned things out to a "T." She did it all herself, and her not being there was...sad. It was a happy time, but it was really, really sad because she was not there. But they carried it off in her honor and it was nice. She would have been happy with how the other teachers carried it off...in her memory. Her plays, the 5th grade play—oh my gosh! For someone to come up with the things she came up with. I mean who would ever think of...? They did "Thriller," and the girls danced with the heels on. The steps, and the music, I mean the time she put in, you know, you just...each year it was something different—it was amazing. It was like where is she getting all of this from? Beautiful voice—she could sing, and she would dance—she could do it all. It's just unbelievable. The graduation dresses—they all looked good, and she made sure all the children had shoes, slips, a tie, a shirt—everything. She was proud of her children. She didn't brag on what she did—she knew what she was doing, and she knew she had it, and it just spoke for itself. It was always positive. She would not leave sometimes until the school is ready to lock-down. She would be here doing something for the children. She would stay and work with them on her time. That's something she didn't have to do—it takes a dedicated person to do that, and that she was. She really was. [My children] learned a lot. Some things they are going to forget, but a lot of it they will always remember. They're going to always remember Ms. Merriex. If they don't remember anybody else, they'll remember her. When she left this world, she left a great part of her behind in all of her children.

Gloria's commitment to her students and community left an imprint that endures. As these stories indicate, a dedicated, culturally responsive teacher can truly empower students to become successful mathematicians while affecting the culture of a school.

Chapter 6

> *I think* ***I'm going by reputation*** *too. They come in the first day of school and just look up at me not knowing what's gonna happen. They hear it through the grapevine where they understand...they think I'm mean, but I told them I don't care what they think. I'm here to teach you. If you're ready to learn, then fine, if not, then the door is always open for you. You know, I don't give them too many options.*
>
> -Gloria Jean Merriex

As we look to the future of mathematics education research and classroom practice, we can glean many ideas from the work and life of Gloria Merriex. There are implications of her work for various areas of mathematics education such as classroom practice and mathematics teaching, mathematics teacher education programs, and future research in the field of mathematics education.

Implications for Classroom Practice

1. Culturally responsive mathematics teachers know their students. These teachers take time to get to know student cultures at all levels, and use this information in their instructional planning and in the course of teaching. Specifically, these educators understand the community culture (e.g. communication patterns, shared values) in addition to what is commonly referred to as "kid culture" (e.g. how 5th graders think and feel, where they are emotionally and intellectually) and relevant pop culture (e.g. what types of music students listen to, who are the big names in entertainment). This information comes from the students themselves, teacher involvement in the community, and a vested interest in students' ways of living. Further, learning about students is not an overnight process; it takes years of personal involvement to build relationships that will allow a teacher to access this type of knowledge, particularly if he or she does not live in the neighborhood surrounding the school.
2. Culturally responsive mathematics teachers are supported by a particular belief system. Effective teachers believe that their students, no matter what statistics say, have the ability to be talented and gifted. Their job is to work to find ways in which students can express their talents while accessing mainstream knowledge. These beliefs are fundamental in providing effective non-verbal communication and building relationships; as teachers we must realize that we are constantly communicating with students even if we don't realize that we are doing so. If a teacher does not truly hold positive beliefs about her population of students, on the other hand, she will communicate these non-verbally and will struggle to build relationships. Students can see straight through

words to what is underneath, and that is what is most important. Lastly, effective teachers are aware of what they believe and continuously reflect and act on those beliefs.

3. Culturally responsive mathematics teachers do not ask students to abandon their home culture. As simple as this might sound, it is fundamental to the success of underrepresented students. Many times, these students are asked to learn from traditional, individualistic, lecture-type instruction, and are deemed disabled if this type of instruction does not provide them with a connection to the material. Culturally responsive teachers are able to teach content, even mainstream content, without asking students to live, act, or learn differently than they would at home. This creates cultural congruity for students, and greatly increases their chances for success. Further, it allows for racial and cultural identity development in the course of teaching, and provides a safe context from which they can learn about societal pressures and misconceptions. Such an environment provides a solid foundation from which students can achieve academic success within their own cultural norms.
4. Constant reflection and revision is necessary for culturally responsive mathematics teaching. Children need to be challenged and engaged in mathematics. In refusing to change particular aspects of instruction, a teacher is not attending to the needs of her students. Reflection allows an educator to see what it is that needs revision while encouraging innovation and change. This reflection-revision process is cyclical and should happen at the macro (i.e. reflecting on entire lessons, units, or ways of teaching) and the micro (i.e. in the course of teaching on a moment to moment basis) levels.
5. Culturally responsive mathematics teachers work beyond the classroom. Specifically, these teachers provide resources to the community in proportion to their personal expectations. For example, if a teacher expects her students to receive help at home, she recognizes that parents might not be familiar with the material, and provides a class that will help them to help their children. If a teacher expects a student to have supplies, she provides them with the necessary materials so that each child has the same opportunity. This type of commitment is a hallmark of successful teaching, and stems from a teacher's belief system, knowledge of students, and relationships.

Implications for Mathematics Teacher Education Programs

1. Multicultural mathematics curricula must be accessible and prevalent throughout teacher education programs. Too often, suggestions for being culturally responsive are tacked on to a curriculum rather than integrated throughout a program. By simply adding a multicultural piece to our existing programs, we are communicating to pre-service teachers that culture is something extra we have to worry about rather than a re-

source that should be used in instruction.

2. Successful examples of culturally responsive mathematics teachers should be touted. We should use teachers who teach with a focus on culture as positive examples who enact the beliefs that we preach in our programs. Further, more care should be taken when assigning pre-service teachers to classrooms for student teaching experiences. By placing these students in traditional classrooms that are based Euro-centric norms, we are undoing all of the good that might have been done by multicultural coursework. Student teachers are likely to mold themselves to their mentor teachers (at least somewhat), so teacher educators must work to find such models.
3. Mathematics teacher educators should model and embody CRT. It is one thing to preach the theory, but quite another to enact it in the class-room. Again, we continuously communicate to our students in verbal and non-verbal ways. If we are saying one thing and doing another (e.g. saying to get to know student culture but keeping a safe distance from our own students), our values lose credibility. If we enact CRT in our own classrooms, however, students are more likely to adopt such practices in their own teaching.

Implications for Future Research

1. Successful models of CRMT should be studied further. In the current literature, there are very few classroom examples of CRT in the context of mathematics. In theorizing about such models, we attain a certain level of understanding, but in discussing what we see in action we are able to more readily understand the relationship to practice. Further, such models can provide us with more salient definitions of CRMT while informing our teacher education programs. In turn, successful multicultural teacher education programs should also be studied and used as models.
2. Student voices should be included in the literature. The effects of CRMT on students' academic achievement, social interactions, and long-term self-concept should be studied. This will help us to more easily evidence the effects of CRMT on students, parents, and communities. Further, it will allow underrepresented student voices to infuse the literature base in the field. If our ultimate goal is improved student achievement in particular populations, it is imperative that we include such voices.
3. Mathematics teacher beliefs should be examined in terms of student outcomes. This link seems to be strong and should be examined as such. This would be another avenue through which student voices could be used, as they should be consulted about such beliefs. It would be interesting to look also at the ways in which teachers' beliefs are developed. Examining the ways in which teacher education programs, for example, shape such beliefs, and how these are then enacted in the

classroom.

4. State and national accountability programs should be studied in relation to cultural bias. What effects do these systems, which deem particular topics to be the most important knowledge to have, inhibit or encourage CRMT? Further, we should examine the effects of scripted curricula, many of which arise from state standards documents, on classroom practice, particularly in high-poverty schools and schools with large populations of students of color.

It is important to remember that while there is no formula for CRMT, we can look to successful models like Gloria Merriex to guide our decisions in teaching mathematics. Through consistent examination of our own beliefs, a commitment to our students, and a willingness to constantly reflect and revise, we can provide each student with equal access to mathematics knowledge.

Chapter 7

Orchestrating Greatness— A Tribute to Gloria Jean Merriex

Dr. Don Pemberton
Director, University of Florida Lastinger Center for Learning

First time visitors to Gloria Merriex's classroom quickly discovered that they were witnessing a gifted maestro at work.

Tall, lean and graceful with a confident countenance, Merriex conducted her 5th grade math classroom at Duval Elementary Fine Arts Academy on the east-side of Gainesville as if it was a great orchestra. Towering over her young protégés, the master teacher worked off of a script that was deeply embedded in her brain but was unknown and unseen to a casual observer.

With eyes firmly fixed on their teacher, the students were led through a dazzling array of exercises and activities that had little resemblance to any other math classroom in the world. Singing, dancing, reciting, writing and constantly moving, Merriex's pupils unraveled the mystery of mathematics.

Merriex, a gifted pedagogue, ingeniously merged music, movement and math into a brilliant mosaic that immersed her students in an exciting new world of learning where the goal was total and absolute mastery of mathematics.

A typical classroom lesson might start with a group sing-along to a hip hop song Merriex had composed that contained essential math facts and formulas. Next, her students might pantomime geometric symbols, followed by an original dance, also choreographed by Merriex that provided visual representations of math facts and formulas. Merriex, always one step ahead of her students, would quickly change pace, returning the pupils to their seats to work on their exercises in their journals. Soon, they were on their feet again, writing their math problems on the board and explaining their reasoning to the class.

Counterintuitive and unorthodox, Merriex believed in teaching the most complex and demanding mathematics principles first, then slowly and methodically adding new skills and concepts. Every day she circled back to what had been learned and taught since the first day of school, always vigilant for slippage and gaps in learning.

By the time the Florida Comprehensive Achievement Test (FCAT) appeared at their doorstep, her students were ready to conquer it. Their confidence was well-placed. Seldom did they disappoint. Year in and year out, her classes had some of the highest FCAT math scores in the state. In the last year that Gloria's students were tested, they achieved the greatest math gains of any 5th grade class in Florida. Recognized for her prowess in teaching math, Merriex was equally successful in helping her students master reading. Year end and year out, Gloria's students aced the FCAT reading test as well.

Merriex's results were even more extraordinary when viewed from the context of the under-resourced community in which her students lived. Keenly aware that more than 95% of her students came from low income homes, she never saw them as poor. Instead, she thought they were rich in potential and was determined to mine that potential and turn it into gold.

In time, word got out about the talented teacher. It was no surprise to her colleagues when she won the teacher of the year award for her school system. Soon the outside world discovered the amazing teacher from Gainesville. Scholars, principals, doctoral students and leaders of philanthropic foundations from around the state and country visited her at work.

She and her students became much in demand conference keynote presenters. Merriex took her class to demonstrate their math prowess, in lively and entertaining performances, that showcased her teaching strategies to audiences in Miami, Tampa, Atlanta and Orlando. They never failed to receive standing ovations.

University of Florida College of Education researchers wrote extensively about the secrets to Gloria Merriex's success. Not only had Merriex created an innovative curriculum and utilized cutting-edge teaching strategies, she also deeply connected with her students' culture, community and aspirations. That connection led to evening math classes for parents, Saturday and summer classes for the children, and incorporating the student's everyday life into the math lessons. Most importantly, she empowered her students, giving them confidence to tackle the most daunting intellectual challenges.

Suddenly and sadly, it all came to an end when Gloria Merriex, fondly known as Jean to her family and loved ones, suffered a massive brain hemorrhage and died.

When she passed away, Gloria Jean Merriex was on the verge of national acclaim. She had recently been awarded a prestigious grant from the W. K. Kellogg Foundation, one of the foremost philanthropic entities in the world. Two days prior to her death she met with collaborators at the University of Florida Lastinger Center for Learning to discuss the Kellogg Foundation project that would share her work with a state and national audience by publishing her curriculum, developing training seminars, underwriting an extensive tour schedule and laying the groundwork for a documentary about her life and practice.

In a separate grant, the Smallwood Foundation awarded funding to the University of Florida Digital Worlds Institute to install high tech equipment and cameras in her classroom so that her teaching could be beamed out to classrooms, researchers and teachers around the world.

Although fame was never her goal, the release in early 2011 of Discovering Gloria, a feature documentary about Merriex's life and success as a teacher, may yet gain the national recognition that she so richly deserved. Directed by Boaz Dvir, and funded by the Kellogg Foundation and the UF Lastinger Center, Discovering Gloria utilizes archive footage and interviews to weave a compelling story of the creative power of a teacher to develop innovative approaches to instill in her students a love of love of learning and a mastery of mathematics.

As word got out of Merriex's death, overwhelming grief overtook her friends, family and fans in Gainesville and around the state and country. All who knew her were deeply affected by their loss, particularly since she was just months away from achieving the national recognition that she so richly deserved.

But like most great teachers, fame was never Merriex's goal. From her first day of teaching to her last day on earth, this virtuoso teacher wanted to orchestrate a life of immense goodness and leave a legacy that would live on in the hearts and minds of her students. Indeed, Merriex leaves a powerful mark on this world. In her students, Merriex instilled a love of learning and confidence in their innate abilities. To her fellow educators, she exemplified the power of teachers to transform lives.

Appendix - Formal Methodology

Research Questions

Given the need for successful models of African American learners, I chose to focus on bringing clarity to readers through an in-depth investigation of one successful teacher's practices in a high-poverty, high-minority school. Based on previous observations, I viewed this teacher as a successful model of CRT, her practices specific to mathematics instruction. As such, the research questions were:

1. How do effective mathematics teachers working in a high-poverty, predominantly African American school context:
    a. Structure instructional practices and interactions?
    b. Establish a learning environment that results in mathematical success?
2. What are the interactions between these phenomena?
3. What part (if any) does culture play in these phenomena?

The current study is a case study meant to shed light on the above questions. Certainly, more teachers are to be included in the refinement of the resulting theory in the future.

Initial Study

The given framework is helpful in constructing ideas of what it means to be a culturally responsive educator. This work, however, was done within the theoretical context of CRT provided by scholars in the field (Gay 2000, Ladson-Billings 1994, for example). The intention of this work was to reveal specific characteristics of CRMT within the larger, more established context of CRT. As such, I fully immersed myself in classrooms of successful mathematics teachers in low income, largely African American communities and schools.

Nomination Process

In doing this research, I did not want to impose my beliefs and perceptions on the community within which I was working; rather, I hoped to obtain knowledge from individuals involved in the neighborhood. I first turned to this community when seeking information about successful teachers, asking them to identify appropriate educators for the study rather than relying on set standards that have been dictated by education literature. Informal conversations in church and school settings, in addition to written correspondence, began my contact with parents and community leaders, a communication pattern which continued throughout the study.

Grounded Theory

The lack of existing data in this area, in addition to the import of the social aspects of culturally responsive practice drove my decision to utilize grounded theory as the driving methodology. When using grounded theory, the researcher is not limited in what he or she can consider data. Rather, all parts of practice, including interactions between individuals, verbal and body language, various types of language patterns, and culturally dictated behaviors were considered data in addition to formal interviews, observations, and artifact collection.

Further, grounded theory employs theoretical sampling, a process by which the researcher uses previously gathered data to drive subsequent data collection activities. This meant that my initial observations and interviews with Gloria helped me to construct questions for subsequent interviews and foci for future observations. Moreover, all data collected in her classroom eventually informed my interviews with those who contributed to this manuscript. Symbolic interactionism, which emphasizes the import of language, communication, and belief systems, was a complimentary theoretical perspective.

Given the systematic process inherent in the use of theoretical sampling, data collection and analysis occurred simultaneously, and findings were corroborated continuously using constant comparison. After my first interview with Gloria, for example, I immediately transcribed, coded, and analyzed the interview. These results were compared to observation data that were being collected, artifacts that were being analyzed, and other data that was coming into consideration. Subsequent data were compared similarly to ensure reliability and to unearth major themes that were recurring.

Member Checks

As I conducted the initial research and eventually wrote this book, I continually went back to available participants to share my work. I did this for several reasons. First, I wanted to ensure that I had represented their words appropriately and in the proper contexts. Second, it was important that the book tap into the interests of those who knew Gloria. As such, I asked them to validate the breadth and depth of the information. Lastly, I wanted to ensure that the message that I had intended was indeed being communicated effectively. To this end, I asked participants and scholars in the field to read parts or all of this work and state what they believed the message to be. This led to re-writes and modifications, but I feel that it added a key element to the overall manuscript.

Sample Research Questions

Interview #1 (Sample Questions):

1. Tell me about your background. What was school like for you? When did you decide that you wanted to be a teacher? Where were you prepared to become a teacher?
2. How would you describe your way of teaching? Have you always taught this way?
3. Does your role in the community influence your teaching? How do you use what you know about students in the course of teaching?
4. What types of teaching methods have given you the most success with this population? As a group, what do you feel it is that they need from a teacher?
5. How do you handle the various ability levels of children in your classroom?
6. Can you tell me about a student who is struggling or has struggled in your class? How did you handle it?
7. How do you handle discipline in your classroom? What do you consider to be misbehavior in the classroom?
8. What types of support (from administration, parents, etc.) are key to your success as a teacher?

Interview #2 (Sample Questions):

1. Can you tell me more about the Math Team?
2. Tell me more about the Evening of Elegance and the 5th grade Graduation. How do rites of passage events like these benefit kids?
3. You talked about cultural influences during our last interview. How does being in your class help students develop racially?
4. When a student acts up in your class or doesn't participate, you sometimes take

them out of the room, might have them call home, and always address them directly with a particular tone. Why does this work with this group of students?
5. You communicate with students in very specific ways. Can you talk about that?
6. You mentioned that as a product of this community, you understand where students are coming from. In what other ways does your background influence your teaching?

Interview #3 (Sample Questions):
1. Last week a new student came into your classroom and was almost immediately sent out. Can you talk about why you handled the situation in this particular way?
2. You mentioned that you won't teach your children to act white. What did you mean by this? How does that relate to your students' racial identity development?
3. You have said several times that you believe that your teaching has an impact on generations and the entire community. How do you feel that the successes at Duval have transformed the community?
4. What are the main outcomes of your teaching?
5. How can we better prepare teachers to work in schools like Duval?

Notes

Preface

1. Definition from Merriam-Webster's Dictionary

Chapter 1

1. All Gainesville, FL demographic information is based on The Census Bureau's 2005-2007 Community Survey 3-year Estimates (retrieved from www.census.gov)
2. This information was retrieved through the School Board of Alachua County website and community relations staff: www.sbac.edu
3. Demographic data retrieved from www.publicschoolreview.com and www.fldoe.org
4. Data retrieved from the School Board of Alachua County website (www.sbac.edu)
5. Adequate yearly progress (AYP) is one of the main components of No Child Left Behind (NCLB). The goal of AYP is to document steady student growth on standardized tests so that 100% of students are proficient in each tested area by 2014. The student body must show AYP as a whole and by disaggregated (by race, disabilities, limited English proficiency, and socioeconomic status) population groups.

Chapter 2

1. A pseudonym has been used
2. A pseudonym has been used

Chapter 3

1. See http://www.ed.gov/nclb/landing.jhtml for more information on NCLB
2. Several analyses (see Lee 2006, for example) have used data gathered from NCLB-driven tests to show that the act has not had an impact on students' mathematics achievement or on the achievement gap.
3. All school performance information, quotes, and data in this section were gathered from the national (http://www.ed.gov/nclb/landing.jhtml) and the state (http://www.fldoe. org/) Department of Education websites.
4. ESE stands for Exceptional Student Education. An "ESE" student is usually one who receives special education services from the school.
5. A pseudonym has been used
6. The IB, or International Baccalaureate, Program is similar to an Honors or Advanced Placement Program. It includes a challenging curriculum and is considered a gateway to college.
7. A pseudonym has been used

Chapter 4

1. The term CRT as it is used here is attributed to the work of Geneva Gay, especially as it pertains to her 2000 book, "Culturally Responsive Teaching: Theory, Research, and Practice." The term "culturally relevant teaching" is used by Gloria Ladson-Billings to describe a similar framework.
2. See appendix A for a full explanation of formal methods
3. Pseudonyms have been used for all children's names in this section
4. Haberman (1995) writes that "star teachers" in poor, high-minority schools do not use information about students' home lives for the purpose of blame, but for the purpose of developing relevant instruction.

Chapter 5

1. Pseudonyms have been used for all participants mentioned in this section
2. Another elementary school in Alachua County; a pseudonym has been used

Works Cited

Agee, Jane. 2004. "Negotiating a teaching identity: An African American teacher's struggle to teach in test-driven contexts." *Teachers College Record* 106 (4): 747-774.

Allen, Brenda A., and A.Wade Boykin. 1991. "The influence of contextual factors on Afro-American and Euro-American children's performance: Effects of movement opportunity and music." *International Journal of Psychology* 26 (3): 373-387.

Allen, Brenda A., and A.Wade Boykin. 1992. "African American children and the educational process: Alleviating cultural discontinuity through prescriptive pedagogy." *School Psychology Review* 21 (4): 586-596.

American Association of Colleges of Teacher Education. 1999. *Teacher education pipeline IV: Schools, colleges, and departments of education.* Washington D.C.: AACTE

Ball, Deborah L., and Hyman Bass. 2000. "Interweaving content and pedagogy in teaching and learning to teach: Knowing and using mathematics." In *Multiple perspectives on the teaching and learning of mathematics*, ed. Jo Boaler, 83-104. Westport, CT: Ablex.

Banks, James A. 1991. "A curriculum for empowerment, action, and change." In *Empowerment through multicultural education*, ed. Christine E. Sleeter, 125-141. Albany, NY: State University of New York Press.

Blumer, Herbert. 1969. *Symbolic interactionism: Perspective and method.* Berkeley, CA: University of California Press.

Bondy, Elizabeth, Dorene D. Ross, Caitlin Gallingane, and Elyse Hambacher. 2007. "Creating environments of success and resilience: Culturally responsive classroom management and more." *Urban Education* 42 (4): 326-348.

Boykin, A.Wade. 1986. "The triple quandary and the schooling of African-American children." In *The school achievement of minority children: New perspectives*, ed. U. Neisser , 56-88. Hillsdale, N. J.: Lawrence Erlbaum Associates.

Boykin, A.Wade. 1995. *Culture matters in the psychosocial experiences of African Americans: Some conceptual, process, and practical considerations.* Paper presented at the annual meeting of the American Psychological Association, in New York, NY.

Boykin, A.Wade, and Brenda A. Allen. 2004. "Cultural integrity and schooling outcomes of African American children from low-income backgrounds." In *Rethinking Childhood*, ed. Peter B. Pufall and Richard P. Unsworth, 104-120. Piscataway, NJ: Rutgers University Press.

Charmaz, Kathy. 2006. *Constructing grounded theory: A practical guide through qualitative analysis.* London: Sage Publications.

Crotty, Michael. 1998. *The foundations of social research: Meaning and perspective in the research process.* Thousand Oaks, CA: Sage Publications.

Delpit, Lisa D. 1995. *Other people's children: Cultural conflict in the classroom.* New York: The New Press.

Fordham, Signithia. 1993. "Those loud Black girls: (Black) women, silence, and gender "passing" in the academy." *Anthropology and Education Quarterly* 24 (1): pp. 3-32.

Fordham, Signithia. 1996. *Blacked out: Dilemmas of race, identity, and success at Capital High.* Chicago: University of Chicago Press.

Freire, Paulo. 1998. *Pedagogy of freedom: Ethics, democracy, and civic courage.* Lanham, MD: Rowman and Littlefield.

Gay, Geneva. 2000. *Culturally responsive teaching: Theory, research, and practice.* New York: Teachers College Press.

Glaser, Barney G. 1998. *Doing grounded theory: Issues and discussions.* Mill Valley, CA: The Sociology Press.

Glaser, Barney G. and Anselm L. Strauss. 1967. *The discovery of grounded theory.* Chicago: Aldine.

Haberman, Martin. 1995. *Star teachers of children in poverty.* West Lafayette, IN: Kappa Delta Pi.

hooks, bell. 1994. *Teaching to transgress: Education as the practice of freedom.* New York: Routledge.

Howard, Gary R. 2006. *We can't teach what we don't know: White teachers, multiracial schools* (2nd Ed.). New York: Teachers College Press.

Hudson, Mildred J. and Barbara J. Holmes. 1994. "Missing teachers, impaired communities: The unanticipated consequences of Brown v. Board of Education on the African American teaching force at the precollegiate level." *The Journal of Negro Education* 63 (3): 388-393.

Hurley, Eric A., A.Wade Boykin, and Brenda A. Allen. 2005. "Communal versus individual learning of a math-estimation task: African American children and the culture of learning contexts." *The Journal of Psychology* 139 (6), 513-527.

Irvine, J. 2003. *Educating teachers for diversity: Seeing with a cultural eye.* New York: Teachers College Press.

Jones, M.Gail, Brett D. Jones, and Tracy Y. Hargrove. 2003. *The unintended consequences of high-stakes testing.* Lanham, MD: Rowman & Littlefield.

Ladson-Billings, Gloria. 1994. The *dreamkeepers: Successful teachers of African American Children.* San Francisco: Jossey Bass.

Ladson-Billings, Gloria. 2001. *Crossing over to Canaan: The journey of new teachers in diverse classrooms.* San Francisco: Jossey Bass.

Ladson-Billings, Gloria. 2004. "Landing on the wrong note: The price we paid for Brown." *Educational Researcher* 33 (6): 3-13.

Lee, Jaekyung. 2006. *Tracking achievement gaps and assessing the impact of NCLB on the gaps: An in-depth look into national and state reading and math outcome trends.* Cambridge, MA: The Civil Rights Project at Harvard University.

Merriam-Webster. 2005. *Merriam-Webster's collegiate dictionary* (11th ed.). Springfield, MA: Merriam-Webster.

Nieto, Sonia. 2002. *Language, culture, and teaching: Critical perspectives for a new century.* Mahwah, NJ: Lawrence Erlbaum Associates, Inc.

No Child Left Behind Act of 2001, Pub. L. No. 107-110.

Pang, Valerie O. and Rich Gibson. 2001. "Concepts of democracy and citizenship: Views of African American teachers." *The Social Studies* 92 (6): 260-266.

Ross, Dorene D., Elizabeth Bondy, Caitlin Gallingane, and Elyse Hambacher. 2008. "Promoting academic engagement through insistence: Being a warm demander." *Childhood Education* 84 (3): 142-146.

Snyder, Thomas D. 1999. Digest of education statistics, 1998. Washington, DC: National Center for Education Statistics, U.S. Department of Education.

Spindler, George D. (Ed.). 1987. *Education and cultural process: Anthropological approaches* (2nd ed.). Prospect Heights, IL: Waveland.

Tatum, Beverly D. 1997. *Why are all the black kids sitting together in the cafeteria? And other conversations about race.* New York: Basic Books.

Walker, Vanessa Siddle. 1996. *Their highest potential: An African American school community in the segregated South.* Chapel Hill: The University of North Carolina Press.

Walker, Vanessa Siddle. 2000. "Valued segregated schools for African American children in the South, 1935-1969: A review of common themes and characteristics." *Review of Educational Research* 70 (3): 253-285.

Ware, Franita. 2006. "Warm demander pedagogy." *Urban Education* 41(4): 427-456.

West-Olatunji, Cirecie A., John C. Baker, and Michael Brooks. 2006. "African American adolescent males: Giving voice to their educational experiences." *Multicultural Perspectives* 89 (4): 3-9.

Yasin, Said. 1999. *The supply and demand of elementary and secondary school teachers in the United States.* Washington, D. C.: ERIC Clearinghouse on Teaching and Teacher Education. (ERIC Document Reproduction Service No. ED 436529).

About the Author

Emily Peterek Bonner is currently an Assistant Professor of Curriculum and Instruction with an emphasis in Mathematics Education in the Department of Interdisciplinary Learning and Teaching at the University of Texas at San Antonio. After receiving her B.A. (mathematics) and M.A.T. from Trinity University, Dr. Bonner taught in Houston public schools for several years. It was here that research interests developed. She received her Ph.D. in Mathematics Education from the University of Florida in 2009.

Dr. Bonner greatly enjoys engaging in teacher preparation and professional development, and teaches at the undergraduate and graduate levels. Her research interests include equity in K-12 mathematics classrooms, culturally responsive mathematics teaching, and successful teachers of underserved student populations and students living in poverty. It is Dr. Bonner's great hope that one day we can achieve equity and access for each student in our mathematics classrooms and schools.

Breinigsville, PA USA
27 October 2010
248141BV00004B/3/P

9 780761 853992